OPPOSING VIEWPOINTS SERIES

THE ECOLOGY CONTROVERSY

OPPOSING VIEWPOINTS

GARY E. McCUEN
DAVID L. BENDER
(Editors)

GREENHAVEN PRESS - ANOKA, MINNESOTA 55303

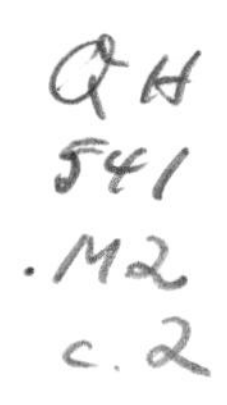

New revised and extended edition

ISBN 0-912616-09-1 Paper Edition
ISBN 0-912616-27-X Cloth Edition

TABLE OF CONTENTS

Page

TABLE OF EXERCISES

A major emphasis of this book is on critical thinking skills. Discussion exercises included after readings are not laborious writing assignments. They are written only to stimulate class discussion and individual critical thinking.

CONFLICTING VALUES: AN INTERNATIONAL PROBLEM

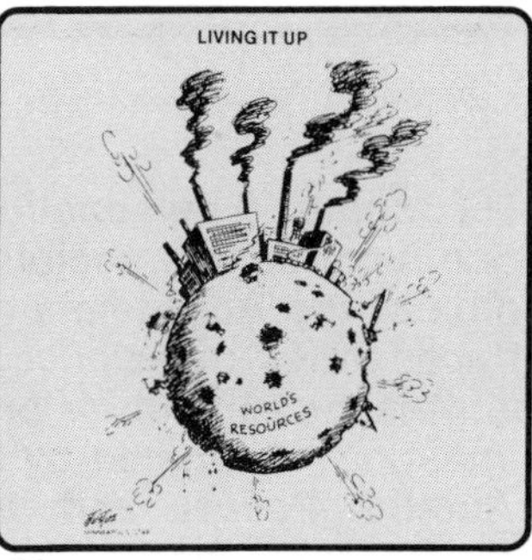

Readings

1. **Limits to Growth**
   The Club of Rome
2. **Third World Declaration on the Human Environment**
   Third World Spokesmen
3. **Technology and Human Values**
   Indira Gandhi
4. **Imperialism and Pollution**
   People's Republic of China

READING NUMBER 1

LIMITS TO GROWTH

by The Club of Rome*

The following reading consists of an interview with Aurelio Peccei, an Italian economist and industrialist. As one of the founders of the Club of Rome, he became its first president.

In April 1968, some 30 personalities from the worlds of industry, science, economics, sociology, government, etc. gathered in Rome at the Accademia dei Lincei, one of the world's oldest academies of science, for an informal discussion on the present and future predicament of man. It was from this meeting, instigated by Italian economist and industrialist Aurelio Peccei and the Organization for Economic Co-operation and Development's Scottish Director-General for Scientific Affairs, Dr. Alexander King, that the Club of Rome was born.

*Excerpted from an interview with the President of the Club of Rome, Aurelio Peccei in **The Limits To Growth, The UNESCO Courier**, January, 1973, pp. 11-12.

Peccei has described the Club as an "invisible college"; it has some seventy members from widely varied backgrounds who share a common conviction that it is urgent to redress the world situation. The Club aims to acquire and spread real understanding of the critical state of human affairs and the uncertain prospects for the future and to propose new policy guidelines for the intelligent management of human affairs.

As a first step the Club commissioned a team of scientists at the Massachusetts Institute of Technology, under the direction of Professor Dennis Meadows, to study the probable dynamics of the world situation with particular attention to the problems of making a deliberate transition from world-wide growth to global dynamic equilibrium . . . They then proceeded to make projections of man's chances of survival in the future. Their ultimate conclusion was that all projections based on growth end in collapse.

This study, the first of a series commissioned by the Club of Rome, was published last year in the form of the now world famous book "The Limits to Growth". The book has aroused enormous controversy (see for instance pages 12,14). In the interview accorded to Unesco recently, extracts of which we publish below, Mr. Aurelio Peccei, President of the Club of Rome, comments on some of the criticisms with which it has been greeted.

QUESTION: What method did the Club of Rome use for such a complex, global study?

Aurelio PECCEI: We based our study on five crucial trends of world concern which, as a starting point, can be said to represent the dynamics, complexities and dangers inherent in the present world system. The first trend is population growth. The second and third are the parallel economic factors of industrial and agricultural growth, in other words, the ability to meet the needs of the growing world population. The fourth factor is pollution, the contamination of the environment with unwanted by-products of industry and agriculture. The fifth is the use we make of our natural resources, the inherited world resources we deplete, all too heedless of the fact that we are living on the capital, not the income.

Q.: Your model comprises five highly complex variables. Did you take into account the variables within these variables?

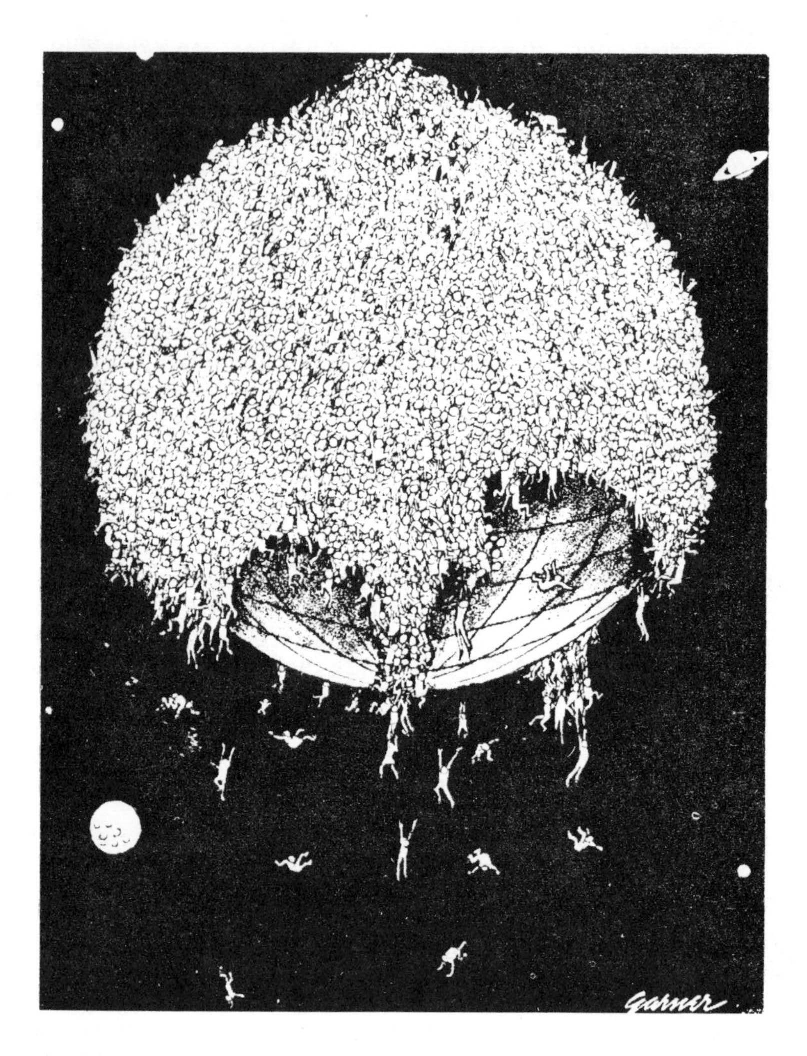

PECCEI: The five variables we chose are interlinked and interact on each other. After trying to take into account all the data that could have a bearing on the interactions, we drew up over a hundred equations whose various curves represent these interrelationships. We fed them into a computer model designed to accept as many world variables as our knowledge or research could identify.

Q.: What global conclusions has the model helped you to reach?

PECCEI: The model is largely indicative in character. Within two to five years we hope to arrive at much firmer conclusions. But the conclusions we can draw already are alarming enough. If present trends continue, exponential growth of production, consumption, pollution and depletion of raw materials will lead to a completely impossible situation: overpopulation of the planet, impoverishment of our environment with our atmosphere and water supplies polluted.

Q.: An optimistic viewpoint suggests that this alarmist attitude is exaggerated.

PECCEI: Our model is purely descriptive of a situation as it exists today and of the possible outcome if present trends continue; the project was not intended as a piece of futurology. The optimists say that the road ahead may seem dangerous but that human ingenuity, science and technology will be capable of solving many of the problems we now face. To my mind, the optimists fail to consider two fundamental factors. The first is the accelerating pace of history; our institutions and our ability to react to problems are not fast enough for us to master these problems in time. Events move more rapidly than we do.

But the second factor is even more fundamental; critical world problems exist for which there are no technical solutions. These are problems of reaction, adaptation and values. Solutions for these problems will have to be sought in the context of social and cultural development. Thus, rather than increase the role of technology in the world, we should, perhaps, attempt to humanize it. This means that we must seek quite different solutions.

Q.: It is claimed that your model corresponds to the situation in industrially developed countries but not to that of the Third World, since it does not include social or political factors.

PECCEI: Such criticisms come closer to the truth. The real need is for a fundamental change in our political and social standpoints. If present trends continue we shall be heading for disaster. Yet we have no new models of the world. We want people to realize, now, that something must be changed in the world. I must stress that the model merely describes the present world situation with all its possibilities and problems. We wanted to learn from our model what this old earth of ours can provide globally and then consider how to make better use of it to eliminate inequalities and tensions. . . .

In a society that will be both fragile and complex, when world population is double what it is now, I fear that computers and other tools that expand the capacity of the human mind will be a necessary aid to life. But if man could regain some essentially human qualities, if injustice were diminished, if wiser generations should succeed us, then perhaps we should have less need of computers to guide us. In short, the choice is between a raising of our ethical standards or an ant-heap existence. As men, I trust, our destiny is not to be mere ants with no higher aspirations than feeding ourselves and achieving material well-being.

CLUB OF ROME CONCLUSIONS

1. If the present growth trends in world population, industrialization, pollution, food production, and resource depletion continue unchanged, the limits to growth on this planet will be reached sometime within the next one hundred years. The most probable result will be a rather sudden and uncontrollable decline in both population and industrial capacity.

2. It is possible to alter these growth trends and to establish a condition of ecological and economic stability that is sustainable far into the future. The state of global equilibrium could be designed so that the basic material needs of each person on earth are satisfied and each person has an equal opportunity to realize his individual human potential.

3. If the world's people decide to strive for this second outcome rather than the first, the sooner they begin working to attain it, the greater will be their chances of success.

The Limits to Growth, A Report For The Club of Rome's Project on the Predicament of Mankind, October 1972.

THIRD WORLD DECLARATION ON THE HUMAN ENVIRONMENT

by Third World Spokesmen*

A group of scientists and others from Asia, Africa, and the Americas, meeting concurrently with the UN Conference on the Human Environment, during June of 1972, issued a statement highly critical of the UN meeting. The statement was also sharply critical of what the group termed "the models of stagnation proposed by certain alarmist Western computer fans."

Use the following questions to assist you in your reading:

1. What is meant by the terms Third World and the Human Environment?
2. What is the cause of the environmental crisis according to the Third World spokesmen?
3. How does overpopulation relate to environmental problems?
4. What do the Third World spokesmen mean when they say technology must be humanized?

*Excerpted from the **Third World Declaration on the Human Environment**, released by the Oi Committee Sweden.

We participants from 41 nations in the Conference on Problems of the Third World and the Human Environment, sponsored by Oi-Committee International, have come together for an independent and critical analysis of the problems of the human environment, parallel to the efforts of the United Nations Conference on the Human Environment in Stockholm June 4-16, 1972. On the basis of prepared papers and discussions in specialized working parties and plenary sessions, we have come to the following understanding.

DECLARATION ON THE THIRD WORLD
AND THE HUMAN ENVIRONMENT

By using the concept THIRD WORLD we focus attention on problems common to peoples who have the shared historical and ongoing experience of being dominated and exploited by other nations. This domination has sharpened the conflict of certain classes ruling over others in our respective societies in the underdeveloped as well as the industrialized regions of the world. The common root of these expressions of domination is in the socio-economic system which allows and favors "development" for one part of society at the expense of another. The price is hunger, disease, environmental deterioration and lack of freedom, lack of access to vital information and of participation.

The HUMAN ENVIRONMENT is the total reality of man's world which includes physical, social, economic, cultural, and political components. We strongly oppose the narrow ecological approach which emphasizes the relations of nature and man, ignoring the fundamental issues of relations between man, societies and classes. Any approach to the problems of the human environment is meaningless unless the economic, social and political inequalities that exist between, as well as within the countries are considered. Our analysis stresses the crucial importance of understanding and solidarity among all oppressed peoples of the world irrespective of where they are living. . . .

Development of the Environment

In each society the social system must be developed based on its own priorities and needs and must integrate all aspects of the human environment. The environmental crisis in both the industrialized countries and the third world is due to the faulty nature of development technology and the social and economic systems that are organized for private gain and the achievement of military power, and not with regard for human needs.

We strongly reject models of stagnation, proposed by certain alarmist Western ecologists, economists, industrialists and computer-fans,[1] and assert that holding economic growth per se responsible for environmental ills amounts to a diversion of attention from the real causes of the problem which lie in the profit-motivation of the systems of production in the capitalist world. Likewise we state that the level of consumption (affluence) per se is not a cause of environmental problems. Therefore the bulk of the so-called ecology movements in most industrialized countries that stress personal ethics of recycling and non-consumption are at best diversionary tactics which fail to put the primary emphasis on the destruction of the profit-oriented system of production.

Agricultural Development

Most specific problem areas in the environment follow the analysis above. For example, the programme of agricultural development known as "the Green Revolution," which is lauded as the answer to the world's food problems, rural unemployment and social stratification, is in reality a package solution involving pesticides, fertilizers, "improved" seeds, irrigation and mechanization, which does irreparable harm to the human environment. This approach has caused severe chemical pollution, elimination of irreplaceable genetic plant materials, increased vulnerability of the agricultural production systems, crop failures, famines, threats to human health from new diseases and malnutrition, and dispossession, dislocation and unemployment of large rural masses as a result of mechanization and the monocultural approach. Although some of the individual elements in this approach may have validity for increasing food production under certain specific conditions, the entire package has, in effect, caused the reverse of its proclaimed aims of social justice. In fact, the widespread introduction of this technique has impeded the essential task of immediate redistribution of resources.

[1] See, for example, D.H. Meadows, et al., **The Limits To Growth.** A Report of the Club of Rome's Project on the Predicament of Mankind. New York, Universe Books, 1972; and *Blueprint for Survival,* **Ecologist**, January 1972.

In summary, far from serving the long-term interests of the rural masses, the so-called Green Revolution has resulted in more profits for the agricultural-business complexes (chemical, machinery, etc.) of the expansionist industrial states. . . .

Population

"We hold that of all things in the world, people are the most precious."[2]

It has often been suggested that the root cause of environmental problems is that the world is "overpopulated"; thus population control is demanded as a solution. We assert, on the contrary, that population growth is neither the most important nor the most decisive factor affecting the human environment. In fact, when population is considered in relation to the development of the environment, there is a vast, as yet underutilized and wasted human energy which can be used to promote development. The question of population is inevitably linked to the question of access to resources. Most arguments for population control are based on the concept of an optimal population size, ignoring that the world is not under one system.

On a global scale, the population problems of the developing countries have arisen primarily since the colonial expansions of the last two centuries, due to the virtual exclusion of the populations of Asia, Africa and Latin America from full access to their own resources. This process of economic exploitation still continues in spite of the nominal independence of various former colonies and dependencies.

We also reject the thesis that the onus of population control must fall on the industrialized countries. Recognizing that in the latter the mode of production of economic goods is a much larger contributor to the environmental crisis than population growth, it is clear that the emphasis must be on changing the modes of production, which, we believe, implies a fundamental change in the socio-economic systems governing the means of production.

[2] Speech by the Chairman of the delegation of the People's Republic of China to the United Nations Conference on the Human Environment, Stockholm, June 1972.

Interpress Film "Attention"

Nevertheless we do not deny that there may be an eventual need to stabilize population growth. This should not be achieved through exerting external pressures of manipulating people to go against their immediate individual interests. Such an approach is intrinsically inhuman and demands a constant technical control over people.

What is needed, in our view, is a process which releases the internal mechanisms by which a population stabilizes itself. The emphasis must be placed on generating a consciousness among people to relate their immediate interests to the broader and long-term interests of the community as a whole. This consciousness, of course, must be preceded by a reorganization of society and the system of production for an equitable distribution of resources.

We therefore strongly condemn the international agencies and aid programs for their involvement in population control policies which are against Third World peoples and which will perpetuate their exploitation. . . .

Self-Determination, National Liberation, War and Weaponry

Modern warfare by the expansionist industrialized states, including the threat of nuclear or biochemical war, presents the greatest urgent single threat to human survival. These states, chief among them the United States of America, are today engaged in their most barbaric drive against the people of the Third World, of which the Indochinese people are the most notable victims. Likewise, the United States is involved in an anti-liberation campaign against the people of Zimbabwe, Namibia, South America, Angola, Mozambique, Guine-Bussau, Palestine, Eritria and others by providing materials to the colonialist wars of Portugal, Israel, Ethiopia, etc. Such domination is vital to the preservation of the U.S. capitalist economy. The armed forces of the United States and other expansionist states are an extension of their policies through which their imperial hegemony is maintained by maneuvering the Third World countries into dependent economic, political and military relationships.

In any discussion concerning the development of the environment, a high priority must be given to the fact that whole cultures and peoples are being decimated, displaced, dispossessed, and in some cases, threatened with imminent extinction.

Many groups who have and wish to maintain their cultural, racial, religious and national identities are victims of "double oppression."[3] In addition to being oppressed in the same way as other exploited peoples, they are persecuted because of their particular ethnic, racial, cultural character. These blatant injustices, particularly against the Indians of North and South America, must be exposed and condemned. We affirm our unequivocal solidarity with their struggles to retain their cultural identity and to defend their right to exist.

We advocate and support the inalienable rights of the peoples of each nation to determine their own socio-economic and political systems. We also believe that when a government is unjust and perpetuates oppression, the oppressed have the right to resort to armed struggles for their liberation movement. We therefore affirm our uncondi-

[3] **Declaration of Barbados**. Symposium on Inter-Ethnic Conflict in South America, January, 1971.

tional support to the liberation movement in Indochina, Africa, and Latin America. We condemn the imperialist and neo-colonialist countries for their naked aggression in these areas.

Humanizing Technology

The carelessness of technology is the product of market-oriented economy where the maximization of profits and price-free social costs are the governing factors. In this manner multinational corporations and national elites are enriched at the cost of environmental deterioration and human degradation. Ecologically faulty technologies such as the production of synthetics (pesticides, inorganic fertilizers, detergents, plastics, synthetic rubber, etc.) can only be justified by market-directed corporations for distorted economic reasons. Processes of this kind have stifled the production of finished natural goods in the Third World, created unparalleled industrial pollution, destroyed socio-economic structures and forced the dependency of the Third World on the expansionist industrialized world.

Technology must be reoriented towards more egalitarian goals to account for the social welfare of peoples and their liberation from oppression by privileged and powerful classes with vested economic interests. Furthermore, the concept of the "transfer of technology" which is supposedly designed for the adaptation of technology to Third World environments is a myth and purely paternalistic in outlook. It fails to consider the social and economic context in which decisions are made regarding the adoption of technologies.

A humane technology for the Third World must necessarily come out of the incentives of the people themselves. This can only happen after a far-reaching social revolution has achieved the goal of total participation by the masses. The new technology must also reinforce many already existing ones such as traditional farming and medical techniques; it must direct innovation in accordance with human needs and environmental imperatives.

Science and Social Responsibility

The new technology must be based on a new concept of science intended only for the social well-being of all peoples. This requirement dictates that for science to benefit the people it must develop out of the struggle for the liberation of mankind. Scientists and other workers alike have the responsibility to participate in the economic and social development of all people and in the struggle for the liberation of mankind as a whole.

We reject the concept of "neutrality of science and education." They can be used to enslave man or to liberate him. Man is a social and historical being and as such has the capacity to change his world of today, which is one of oppression, to the world of tomorrow, which will be one of liberation.

We denounce the exploitative values forced on children through competitive education based on individual achievement. We condemn the attitude of the superiority of "intellectual" over manual labour. Education must enhance self-awareness and social concern and lead to a new consciousness for greater participation of the people at all levels of decision-making.

Finally, we do not believe that the United Nations organization will be able to find solutions to the problems of mankind so long as it is directed and controlled by the very powerful expansionist industrial and military states which oppress the peoples of the exploited world. It is absurd to expect the oppressors to initiate and implement the process of liberation of the oppressed. We assert that the problems of this world can only be solved by the people. Therefore we call on all the people of the world to intensify their struggle against the forces that oppress us.

If the three-fourths of mankind represented by underdeveloped countries were to squander natural resources at the same rate (in per capita terms) as, for example, the United States or the Western European countries, there would not be enough oxygen to go around and there would not be enough metals for industry, while, on the other hand, there would be so much carbon, sulphur, and nitrogen dioxide that mankind would be pushed toward extinction.

Miguel A. Ozorio De Almeida, chairman of the Brazilian delegation to the United Nations Conference on the Human Environment at Stockholm in June 1972.

EXERCISE 1

CAUSE AND EFFECT RELATIONSHIPS

This discussion exercise provides practice in the skill of analyzing cause and effect relationships. Causes of human conflict and social problems are usually very complex. The following statements indicate possible causes for international pollution. Rank them by assigning the number (1) to the most important cause, number (2) to the second most important, and so on until the ranking is finished. Omit any statements you feel are not causative factors. Add any causes you think have been left out. Then discuss and compare your decisions with other class members.

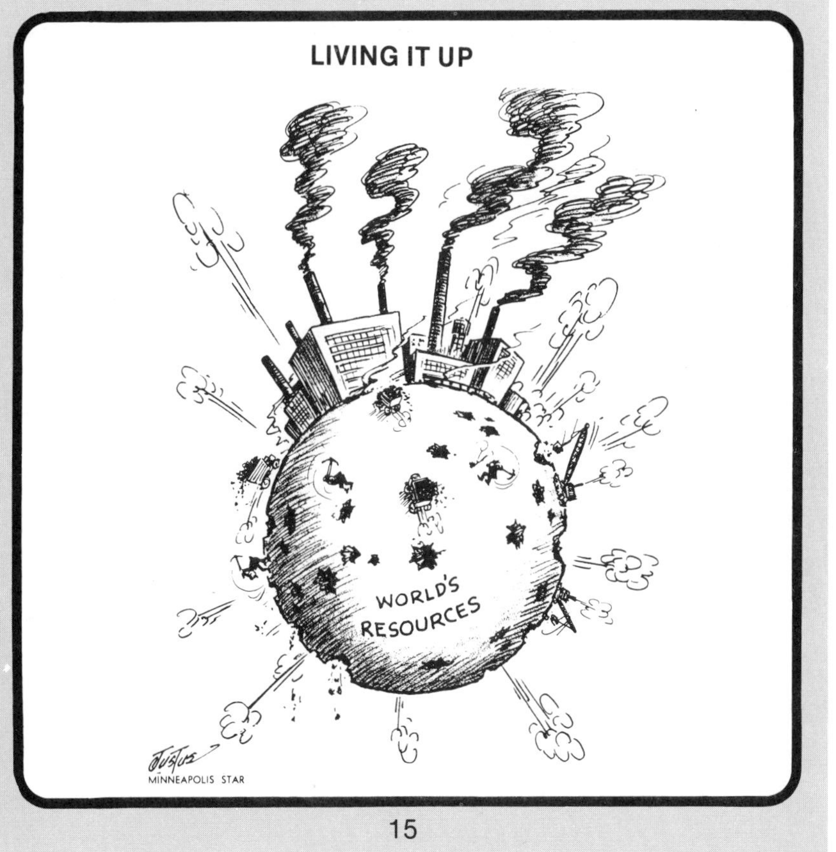

_____ 1. Rapid growth of world population

_____ 2. Expansion of industry in all countries

_____ 3. Lack of environmental concern by industrialists who are motivated by profits

_____ 4. Lack of concern by citizens of the world

_____ 5. The apathy of religious leaders

_____ 6. Lack of action by politicians who fear reprisals from pressure groups

_____ 7. Absence of technology to reduce industrial pollution

_____ 8. Modern warfare

_____ 9. Indifference of affluent consumers in all countries

READING NUMBER 3

TECHNOLOGY AND HUMAN VALUES

by Indira Gandhi*

Indira Gandhi is the Prime Minister of India. She presented the following statement at The United Nations Conference on the Human Environment at Stockholm, Sweden, June 1972.

The following questions should help you examine the reading:

1. How have nations treated the environment?
2. What countries have been the greatest polluters?
3. What part has population played in pollution?
4. What attitude should the poor nations take toward industrial development in their own countries?

*Excerpted from a statement by Indira Gandhi at the United Nations Conference on the Human Environment. The Centre for Economic and Social Information at United Nations European Headquarters, Geneva, Switzerland.

I have had the good fortune of growing up with a sense of kinship with nature in all its manifestations. Birds, plants, stones were companions and, sleeping under the star-strewn sky, I became familiar with the names and movements of the constellations. But my deep interest in this our "only earth" was not for itself but as a fit home for man.

One cannot be truly human and civilized unless one looks upon not only all fellowmen but all creation with the eyes of a friend. . . .

Along with the rest of mankind, we in India . . . have been guilty of wanton disregard for the sources of our sustenance. . . .

It is sad that in country after country, progress should become synonymous with an assault on nature. . . .

Many of the advanced countries of today have reached their present affluence by their domination over other races and countries, the exploitation of their own masses and their own natural resources. They got a head start through sheer ruthlessness, undisturbed by feelings of compassion or by abstract theories of freedom, equality or justice. . . .

The Greatest Polluters

On the one hand the rich look askance at our continuing poverty — on the other they warn us against their own methods. We do not wish to impoverish the environment any further and yet we cannot for a moment forget the grim poverty of large numbers of people. Are not poverty and need the greatest polluters? For instance, unless we are in a position to provide employment and purchasing power for the daily necessities of the tribal people and those who live in or around our jungles, we cannot prevent them from combing the forest for food and livelihood; from poaching and from despoiling the vegetation. When they themselves feel deprived, how can we urge the preservation of animals? How can we speak to those who live in villages and in slums about keeping the oceans, the rivers and the air clean when their own lives are contaminated at the source? The environment cannot be improved in conditions of poverty. Nor can poverty be eradicated without the use of science and technology.

Must there be conflict between technology and a truly better world or between enlightenment of the spirit and a higher standard of living? . . .

It is an oversimplification to blame all the world's problems on increasing population. Countries with but a small fraction of the world population consume the bulk of the world's production of minerals, fossil fuels and so on. Thus we see that when it comes to the depletion of natural resources and environmental pollution, the increase of one inhabitant in an affluent country, at his level of living, is equivalent to an increase of many Asians, Africans or Latin Americans at their current material levels of living.

The inherent conflict is not between conservation and development, but between environment and the reckless exploitation of man and earth in the name of efficiency. . . .

All the "isms" of the modern age — even those which in theory disown the private profit principle — assume that man's cardinal interest is acquisition. The profit motive, individual or collective, seems to overshadow all else. This overriding concern with Self and Today is the basic cause of the ecological crisis.

Pollution is not a technical problem. The fault lies not in science and technology as such but in the sense of values of the contemporary world which ignores the rights of others and is oblivious of the longer perspective.

There are grave misgivings that the discussion on ecology may be designed to distract attention from the problems of war and poverty. We have to prove to the disinherited majority of the world that ecology and conservation will not work against their interest but will bring an improvement in their lives. To withhold technology from them would deprive them of vast resources of energy and knowledge. This is no longer feasible nor will it be acceptable.

Development vs. Environment

The environmental problems of developing countries are not the side effects of excessive industrialization but reflect the inadequacy of development. The rich countries may look upon development as the cause of environmental destruction, but to us it is one of the primary means of improving the environment for living, of providing food,

Does the fact that technological progress and economic growth are at present destroying the Third World's environment justify the halting of growth in these areas, as some people insist? I do not believe so. It seems to me quite absurd to propose a zero growth rate for the Third World when the peoples of these areas consider economic development to be their last hope of emerging from the crushing burden of poverty.

Josue De Castro, President of the International Centre for Development, in Paris.

water, sanitation and shelter, of making the deserts green and the mountains habitable. The research and perseverance of dedicated people have given us an insight which is likely to play an important part in the shaping of our future plans. We see that however much man hankers after material goods, they can never give him full satisfaction. Thus the higher standard of living must be achieved without alienating people from their heritage and without despoiling nature of its beauty, freshness and purity so essential to our lives.

The most urgent and basic question is that of peace. Nothing is so pointless as modern warfare. Nothing destroys so instantly, so completely as the diabolic weapons which not only kill but maim and deform the living and the yet to be born; which poison the land, leaving long trails of ugliness, barrenness and hopeless desolation. What ecological project can survive a war?

Life is one and the world is one, and all these questions are interlinked. The population explosion, poverty, ignorance and disease, the pollution of our surroundings, the stockpiling of nuclear weapons and biological and chemical agents of destruction are all parts of a vicious circle. Each is important and urgent but dealing with them one by one would be wasted effort. . . .

If there is to be a change of heart, a change of direction and methods of functioning, it is not an organization or a country — no matter how well intentioned — which can achieve it. While each country must deal with that aspect of the problem which is most relevant to it, it is obvious that all countries must unite in an overall endeavor. There is no alternative to a cooperative approach on a global scale to the entire spectrum of our problems.

READING NUMBER 4

IMPERIALISM AND POLLUTION

by People's Republic of China*

The following statement was made by Tang Ke. He was the chairman of the delegation of the People's Republic of China at the United Nations Conference on the Human Environment.

As you read try to answer the following questions:

1. Serious pollution is the result of what two main developments?
2. What is the relationship between population and protection of the human environment?
3. What attitude should the poor nations take toward industrial development and pollution in their own countries?

*From a statement by Tang Ke at the United Nations Conference on the Human Environment. The Centre for Economic and Social Information at United Nations European Headquarters Geneva, Switzerland.

In a growing number of areas of the world the human environment is now subjected to contamination and damage, some of which have even become serious social problems. Air has been poisoned, garbage has become a catastrophe, rivers, seas and oceans have been polluted, the growth and reproduction of fauna and flora have been affected, economic development has been hampered, and the health of the broad masses of people has been seriously threatened or harmed. This cannot but arouse the deep concern of the people of all countries. To conserve and improve the human environment, to fight pollution, has become an urgent and vital issue in ensuring the healthy development of the human race. In our opinion, the increasingly serious pollution and damage of the human environment in certain regions, becoming an outstanding issue, is mainly the result of the development of capitalism into imperialism and particularly the policies of plunder, aggression and war frenziedly pursued by the superpowers.

Since World War II, U.S. imperialism, in its attempt for world hegemony, has committed aggression and interference everywhere, particularly their ruthless suppression of the struggles of the people in Asia, Africa and Latin America to win and safeguard national independence. As is known to all, in its war of aggression against Vietnam, Cambodia and Laos, U.S. imperialism, in total disregard of the opposition by the people of the world, including the American people, has not only repeatedly intensified its naval and air attacks against the whole of Vietnam, but has become panic-stricken and has been fanatically using toxic chemicals and poisonous gas continuously in southern Vietnam, Laos and of late even northern Vietnam. These barbarous atrocities on the part of the United States have resulted in the massive killing of innocent old people, women and children as well as unprecedented and serious destruction of the human environment. Innumerable houses have been razed to the ground, great stretches of fertile land have been reduced to bomb craters, rivers and water resources have been polluted, forests and agricultural crops destroyed and certain biological species are faced with the danger of extinction. These shocking atrocities committed by U.S. imperialism cannot but arouse the utmost indignation of the people throughout the world as well as all those who are engaged in the work of protecting the human environment. . .

The delegation of the People's Republic of China holds that our Conference should not remain indifferent towards such atrocities. We should strongly condemn the United

States for their wanton bombing and shellings, use of chemical weapons, massacre of the people, destruction of human lives, annihilation of plants and animals and pollution of the environment.

The immense destruction brought about by indiscriminate bombing, by large scale use of bulldozers and herbicides is an outrage sometimes described as ecocide, which requires urgent international attention. It is shocking that only preliminary discussions of this matter have been possible so far in the United Nations and at the conferences of the International Committee of the Red Cross, where it has been taken up by my country and others. We fear that the active use of these methods is coupled by a passive resistance to discuss them.

Olof Palme, Prime Minister of Sweden at the United Nations Conference on the Human Environment at Stockholm in June 1972.

Here we would also like to deal briefly with the relationship between population growth and protection of the human environment. We hold that of all things in the world, people are the most precious. The masses have boundless creative power. To develop social production and creative social wealth depends on people, and to improve the human environment also depends on people. The history of mankind has proved that the pace of development of production, science and technology always surpasses by far the rate of population growth. The possibility of man's exploitation and utilization of natural resources is inexhaustible. Moreover, with the progress in science and technology, man's use of natural resources will increasingly grow in depth and scope. Mankind will create ever greater quantities of wealth to meet the needs of its own subsistence and development. Mankind's ability to transform the environment will also grow continuously along with social progress and the advance of science and technology. This can be illustrated by the case of China. The rate of population growth in our country has been relatively rapid. While in 1949 our population was over 500 million, in 1970 it exceeded 700 million. However, because we had driven out the imperialist plunderers and overthrown the system of exploitation the standard of living of the people has not lowered but gradually improved in spite of the relatively rapid growth of the population. The country has not become poorer but has gradually become prosperous, the people's living environment has not deteriorated but has gradually improved. Of course, this in no way means that we approve of the unchecked growth of the population. Our government has always advocated family planning, and the publicity, education and other measures adopted over the years have begun to produce some effects. It is wholly groundless to think that population growth in itself will bring about pollution and damage of the environment and give rise to poverty and backwardness. . .

We support the developing countries in building their national economies on the principle of independence, exploiting their natural resources in accordance with their own needs and gradually improving the well-being of their people. Each country has the right to determine its own environment standards and policies in the light of its own conditions, and no country whatsoever should undermine the interests of the developing countries under the pretext of protecting the environment.

EXERCISE 2

REVOLUTION AND CHANGE

The following exercise will explore your attitude toward change. Sometimes change brings progress, other times pain and suffering, and frequently both progress and human suffering are by-products of social, political, scientific, and technological change. Change can occur slowly, or it can come suddenly and quickly (revolutionary change).

Consider each of the following circumstances carefully. Mark (G) whenever you feel gradual change is needed. Mark (R) for circumstances that you believe demand revolutionary change. And mark (S) if you think the status quo should be maintained (no change needed).

G = Gradual Change
R = Revolutionary Change
S = Status Quo

____ 1. The present growth in world population

____ 2. The economic system in the United States

____ 3. Unrestrained worldwide economic expansion

____ 4. Large military budgets in most countries of the world

____ 5. Industrial development in Third World countries

____ 6. A return to organic farming in industrial countries

____ 7. The attitude that man is dominant over the environment and free to exploit it

CHANGING VALUES: A NATIONAL PROBLEM?

Readings

READING NUMBER 5

WESTERN CULTURE AND THE ENVIRONMENT

by Luther P. Gerlach*

Luther P. Gerlach is an Associate Professor of Anthropology at the University of Minnesota. The reading that follows was presented at a conference of the Minnesota Association for Conservation Education.

Consider the following questions while reading:

1. Gerlach claims that western culture's view of man and nature is a cause of the ecology crisis. What points does he make?
2. What kind of revolutionary change does the author say is needed to save the environment?

*Luther P. Gerlach, *The Mobilization of Human Resources to Save Natural Resources: Movement Dynamics in Action.*

Western culture tends to see man not as a part of nature, but rather as apart from and dominant over the natural environment. This view is especially marked among North Americans. The natural environment thus becomes a resource not to be *interacted* with but rather to be *exploited for profit.* Science and technology are developed as the means to accomplish this exploitation. The goal of the vast majority of economic development projects is maximization of various types of financial or material gain and expansion at the expense of the non-human biophysical environment and organisms.

Even the biophysical "health" of man is sacrificed to achieve financial or material gain and growth. The political subsystems and religion in western societies have both supported such strategies of environmental exploitation.

Increasingly, Western man is perceiving that such strategies of exploitation are maladaptive in respect to human health and general welfare. But man finds it difficult to change course and devise new strategies of adaptive resource management. For one, the socio-cultural systems of the Western world have been geared to exploitation and if nothing else, inertia perpetuates the present course of action. Changes will cost money and energy and man must be persuaded to pay this cost. For another, man has simply not developed a new order of goals and priorities against which new and more desirable strategies can be programmed. What happens, for example, when interest groups conflict over the locating of a site for the construction of an airport? One argues that a certain site is best because it maximizes efficient airport construction at lowest cost proposed with good access to urban centers. Another opposes this site because its construction threatens to lower water tables and otherwise damage a nearby wildlife reserve. Another supports this proposed site because it is thought that it will bring in industry and economic growth. And yet another group opposes the site because it believes that an operational airport will destroy the attractiveness of the area as a residential site, and lower property values. Each group is obviously making judgements according to different and opposing goals. Conventional wisdom and traditional goals of economic growth at any cost would support construction of the airport at this site but these approaches are now being challenged. No approach is now clearly dominant, and thus we have no clear basis to construct socially desired or desirable strategies for resource management.

The ecological crisis is, at heart, a religious matter. If that proposition seems too far-fetched, consider the logic: our treatment of the environment is simply the translation into behavior of the values we hold concerning nature. In religious terms, we are acting out what we believe about the creation, that is, what we believe to be the meaning of life. This is a religious effort, even if some of us are not particularly religious people. . . .

The damage we are doing to the environment is partly rooted in confusion arising from our Western religious heritage — what we often call the Judeo-Christian tradition. And this is true even though many of us have broken formal ties with, and may even feel it appropriate to speak against, that tradition. After all, if we share common attitudes toward nature, attitudes which come from a shared thought tradition, how we verbalize them, or whether we are pro or con toward organized churches, makes little difference. The practical result — the abuse and degradation of the environment — is the same. . . .

The roots of both the Jewish and Christian traditions lie in the writings of the ancient Hebrews, whose insights were earthy, pragmatic and possessed of a positive attitude toward physical life. The Hebrews did not conceive of an "afterlife" because they viewed man as a vitalized, physical being rather than as an incarnate spirit. In the language of the creation myth, God breathed his own breath (i.e., vitality) into man. But that did not make man an eternal spirit; rather, he became a "living being" framed of the soil of earth, a creature of dust sharing God's quality of "aliveness."

Hebrew man's notion of redemption also was tied to material existence, and had to do with the quality of physical well being: to be blessed by God was to live at peace with one's neighbors, free from material want and surrounded by healthy children.

When Christianity began to expand into the Graeco-Roman world it encountered Greek philosophy. The Greeks split reality into separate realms, a way of thinking we have come to call "dualism." On the one hand was the physical realm, on the other the realm of essence or spirit. Since physical things are impermanent and waste away in time, they were seen as illusory. In the realm of essence, however, things were viewed as eternal. (Greek art and architecture were attempts to reflect the perfection of the eternal.) As a result, the Greeks negatively valued the physical and positively valued the spiritual. From this it followed the material was "bad," spiritual was "good." . . .

In summary, then, our traditional preoccupation to prepare an immortal soul for an afterlife is alien to the biblical tradition. It results from the application of Greek definitions to Hebrew concepts. What is more, it has caused an irresponsible rejection of a "bad" physical life — as opposed to a "good" spiritual one — rather than a responsible celebration of it. . . .

In our confusion, we have treated the physical earth as a vile thing, not a creature of God's loving purpose, because we have assumed it is our destiny to escape it, not to live in it. Such a view permits us to extinguish whole species of life on the wretchedly flimsy ground that they are of no immediate economic value to us, and therefore have no reason to exist, without consideration for their inherent value as sharers in createdness and life.

Dr. Don C. Skinner, Hamline University, in the **Hamline University Bulletin**, undated.

In addition to environmental exploitation which arises because man wishes to maximize economic and material gain, there is also environmental damage and pollution because man is simply careless, lazy, unconcerned, or destructive. Here again man finds it very difficult to end these maladaptive attitudes and practices which ruin his environment. . . .

Legislation, voluntary association activity, public information and education programs set the stage for more fundamental and broad based change in treatment of the environment, but cannot themselves make such change. They cannot transform the behavior and attitudes of the mass of population at the grass roots. This was also the case in American civil rights activities from WW II to 1965.

Attempts to achieve widespread integration and other civil rights goals through legislation, voluntary association activities, and appeals to public reason and conscience also only set the stage for but did not achieve full scale change.

As yet, those who strive for significant and broad-based change in use of the environment do not have enough real power to achieve such change. Although they have made gains, they do not as yet have enough power to direct change through the conventional political system. Appeals to reason and conscience are also insufficient to generate and motivate the energies of the mass of Americans for major change. Ordinary voluntary associations for collective effort, such as the "Save the River X" group, can do much to promote change, but they have certain significant limitations on this ability.

For one, they lack a process by which their members can become committed to strive in the face of opposition and cost; an all too common complaint voiced by leaders of such associations is that only a few members are really willing to sacrifice time, effort, and money for the cause.

For another, and partly because of insufficient commitment, they lack the ability to spread out rapidly at the grass roots, expanding membership among people of all socio-economic class and background. Thus, they are unable to change the pervasive and deep-seated popular attitudes and practices which contribute importantly to, or cause the very problems of, environmental abuse which must be overcome.

How to Achieve Fundamental Change

Fundamental change in personal attitudes and socio-cultural institutions can only come about through a mass social movement, or more precisely a *movement of personal transformation and revolutionary change.* The black power movement and associated movements of response in white society, especially in the church and university, are changing this country to achieve a more just, moral and adaptive use and treatment of human resources. By the same token it will take a movement of personal transformation and socio-cultural change to achieve the fundamental changes in perception, utilization, and treatment of man's biophysical resources. Through such a movement, human resources can be mobilized and directed to establish among man a more adaptive interrelationship with environment. It will require the energies of a movement to rethink, revise concepts, and then change goals and strategies for maximization. It will require the energies of a movement to secure the appropriate political and economic power to accomplish objectives, and to provide the ideological and "religious" dimension to sanction and motivate such change.

In other words, only a movement of revolutionary change can reshape the human component in the human ecosystem sufficiently to save the ecosystem.

BELOW OLYMPUS

"Yes, sir, we enjoy the highest standard of living known to the world!"

READING NUMBER 6

ECOLOGY CRISIS: CRIES OF WOLF

by A.L. Jones*

Dr. A.L. Jones is a research scientist at Standard Oil of Ohio. This reading was originally delivered as a speech to the Public Relations Society of America, Greater Cleveland Chapter, Hotel Statler Hilton, Cleveland, Ohio, December 14, 1971.

As you read consider the following questions:

1. How is oxygen manufactured in the atmosphere? Why is this significant?
2. What effect does carbon monoxide, produced by automobile exhaust, have on nature?
3. How should DDT be used in the future?
4. What does mankind have to do with the disappearance of animal species?

*A.L. Jones, *A Question of Ecology*, **Vital Speeches**, April 1, 1972, pp. 381-83. Reprinted with permission from **Vital Speeches.**

For several years I have been deeply concerned about reports of the destruction of our environment as a result of technological recklessness, overpopulation and a religious and philosophical outlook that gives little consideration to the preservation of nature. My studies in this area of concern have turned up evidence that I feel compelled to share with you. I welcome this opportunity to do it.

I find that many people I talk to are filled with gloom and believe we have no future. Some of them blame our apparent demise on the Judeo-Christian ethic that it is God's will that man exploit nature for his proper ends and that we have overdone it. Others recommend that we return 2500 years and embrace the practices of druidism. Many express a disdain for science and mistrust technology in general.

Our automobiles are no longer a wondrous method of freeing man from his immobility but instead have become terrible polluters and ultimately piles of junk to desecrate the landscape. Electricity, which has been the most convenient form of energy ever available, has come into disrepute. The industry that produces it is looked upon as an evil organization of the establishment whose objective is to create new radiation hazards with atomic plants, cut down trees, stick poles in the ground and pump smoke into the air to poison all of us.

Many believe we are seriously depleting the oxygen of the atmosphere and replacing it with toxic substances such as carbon monoxide. Some say that Lake Erie is dead and all of us will be next. It is a gloomy picture being painted. This outlook is not justified by the evidence I have been able to find. . . .

Throughout my formal education I have been taught that oxygen in our atmosphere is supplied by green plants using the process of photosynthesis. It is known that plants take in carbon dioxide and, through activation by sunlight, combine it with water to make starches and cellulose and give off oxygen. In this way the whole chain of plant and animal life is sustained by energy from the sun. When the vegetable or animal materials thus produced are eaten, burned, or allowed to decay they combine with oxygen and return to the carbon dioxide and water from whence they came. We all know this. What is the surprise?

The surprise is that most of the oxygen in the atmosphere doesn't come from photosynthesis. The evidence is now overwhelming that photosynthesis is quite inadequate to have produced the amount of oxygen that is present in our atmosphere. The reason is that the amount of oxygen produced by photosynthesis is just exactly enough to convert the plant tissue back to the carbon dioxide and water from which it came. In other words, the net gain in oxygen due to photosynthesiss is extremely small. The oxygen of the atmosphere had to come from another source. A most likely possibility involves the photodissociation of water vapor in the upper atmosphere by high energy rays from the sun and by cosmic rays. This process alone could have produced, over the history of the earth (4.5×10^9 years), about seven times the present mass of oxygen in the atmosphere. . . .

The significance of this information is that the supply of oxygen in the atmosphere is virtually unlimited. It is not threatened by man's activities in any significant way. If all of the organic material on earth were oxidized it would reduce the atmospheric concentration of oxygen by less than 1 percent. We can forget the depletion of oxygen of the atmosphere and get on with the solution of more serious problems. . . .

As many of you know, the most toxic component of automobile exhaust is carbon monoxide. Each year mankind adds over two hundred million tons of carbon monoxide to the atmosphere. Most of this comes from automobiles. Until this year I had been concerned about the accumulation of this toxic material because I use it daily in my research and know that it has a life in dry air of about 3 years. For the past several years, monitoring stations on land and sea have been measuring the carbon monoxide content of the atmosphere. Since the ratio of automobiles in the northern and southern hemispheres is 9:1 respectively, it was expected that the northern hemisphere would have a much higher concentration of atmospheric CO. Measurements show that there is no difference in CO amounts between the hemispheres and that the overall concentration in the air is not increasing at all.

Early in 1971, scientists at the Stanford Research Institute in Palo Alto disclosed that they had run some experiments in smog chambers containing soil. They reported that carbon monoxide rapidly disappeared from the chamber. They next sterilized the soil and found that now the carbon monoxide did not disappear. They quickly identified the organisms responsible for CO removal to be fungi of the aspergillus (bread mold) and penicillin types. These organisms, on a world wide basis, are using all of the

200 million tons of CO made by man for their own metabolism, thus enriching the soils of the forests and the fields.

It is commonly believed that pollutants render the air unfit. False. Without salt or dust — both pollutants — we would not have rainfall.

Another pollutant is a form of oxygen known as ozone. At ground level it can damage vegetation and cause rubber to crack. But in the upper atmosphere ozone helps filter out radiation from the sun. Without it, life could not exist on earth. Everything would burn to a crisp.

On a delightful walk through an aromatic pine forest, you're breathing a type of hydrocarbon called terpenes. They're pollutants, too.

The haze which makes the Blue Ridge Mountains look blue from a distance is the result of these terpene hydrocarbons reacting with sunlight. . . .

Not everything discharged from an automobile engine or power plant is harmful. In fact, some of the pollutants are very useful. For example, carbon dioxide. Vegetables depend on carbon dioxide for survival.

Plants, in effect, convert man's waste back into products essential for man's survival.

Even the noxious-looking fumes from a diesel-engined bus or truck, while offensive in appearance and odor, are regarded as nondangerous, little more than a nuisance.

From *Myths About Pollution*, an editorial in the September 1968 issue of **Nation's Business**.

This does not say that carbon monoxide is any less toxic to man. It does say that, in spite of man's activities, this material will never build up in the atmosphere to dangerous levels except on a localized basis. . . .

Newspapers have published statements by Norman E. Borlaug, Nobel Peace Prize winner, on his opposition to the banning of DDT. Dr. Borlaug is a competent scientist. He won the Nobel prize because he was able to develop a new strain of wheat that can double the food production per acre anywhere it can be grown.

Dr. Borlaug said "If DDT is banned by the United States, I have wasted my life's work. I have dedicated myself to finding better methods of feeding the world's starving populations. Without DDT and other important agricultural chemicals, our goals are simply unattainable."

My investigations into this matter strongly verify the statements of Dr. Borlaug. I find that DDT has had a miraculous impact on arresting insect borne diseases and increasing grain production from fields once ravaged by insects. According to the World Health Organization, malaria fatalities alone dropped from 4 million a year in the 1930's to less than 1 million per year in 1968. Other insect borne diseases such as encephalitis, yellow fever and typhus fever showed similar declines. It has been estimated that 100 million human beings who would have died of these afflictions are alive today because of DDT.

DDT and other chlorinated compounds are supposedly endangering bird species by thinning of the egg shells. I am not sure this is true. The experiments I found concerning this were not conducted in such a manner that positive conclusions could be drawn from them. Even if it is true, I believe that the desirable properties of DDT so greatly outnumber the undesirable ones that it might prove to be a serious mistake to ban entirely this remarkable chemical.

Many people feel that mankind is responsible for the disappearance of animal species. I find that in some instances man may hasten the disappearance of certain species. However, the abundance of evidence indicates that he has little to do with it. About 50 species are expected to disappear during this century. But it is also true that 50 species became extinct last century and the century before that. Dr. T.H. Jukes of the University of California points out that about 100 million species of animal life have become extinct since life began on this planet about 3 billion years ago. Animals come and animals disappear. This is the essence of evolution as Mr. Darwin pointed out many years ago. Mankind is a relatively recent visitor here. He has had nothing to do with the disappearance of millions of species that preceded him.

It is of interest to note that man has not been successful in eliminating a single insect species, in spite of his all-out war on certain undesirable ones in recent years, to the best of my knowledge.

For those who wish to return to the "good old days" when we didn't have dirty industries and automobiles to pollute the air, let's consider what life was really like in America one hundred and fifty years ago. For one thing, life was very brief. The life expectancy for males was 38 years. It was a gruelling 38 years. The work week was 72 hours. The average pay was $300 per year. The women's lot was even worse. They worked 98 hours a week, scrubbing floors, making and washing clothes by hand, bringing in firewood, cooking in heavy iron pots and fighting off insects without screens or pesticides. Most of the clothes were rags by present day standards. There were no fresh vegetables in winter. Vitamin deficiency diseases were prevalent. Homes were cold in winter and sweltering in summer.

Every year an epidemic could be expected and chances were high that it would carry off someone in your family. If you think that water pollution is bad now, it was more deadly then. In 1793, one person in every five in the city of Philadelphia died in a single epidemic of typhoid fever as a result of polluted water. Many people of that time never heard a symphony orchestra or traveled more than twenty miles from their birthplace during their entire lives. . . . I wonder how many informed people want to return to the "paradise" of one hundred and fifty years ago. Perhaps the simple life is not so simple.

Many of us are alarmed by the dire announcements from technically untrained people and from scientists who have not bothered to check their assumptions against the evidence. We have gone off half-cocked with expensive measures in some cases to solve problems that are more imaginary than real.

DETERMINING PRIORITIES

EXERCISE 3

Pretend that you are the Secretary of the Treasury and the President has asked you for a recommendation as to how much money should be spent in the coming fiscal year on the following national problems. Assuming you have a National Budget of 240 billion dollars to work with, how much money would you advise the President to spend on each of the following problems?

CITIZEN ON A TIGHTROPE

$____ 1. National Defense

____ 2. Foreign Aid

____ 3. Space Research

____ 4. Agricultural & Rural Development

____ 5. Urban Development and Housing

____ 6. Commerce and Transportation

____ 7. Education

____ 8. Health

____ 9. Income Security: Social Security and Unemployment Compensation

____ 10. Veterans' Benefits and Services

____ 11. Pollution Control

____ 12. Energy Research

THE GROSS NATIONAL PRODUCT AND POLLUTION

by Joseph W. Krutch*

Joseph W. Krutch, who recently passed away, was the author of many books, several of which dealt with nature topics. Among his many other accomplishments, he was an associate editor of **The Nation**, a professor of English, and the recipient of many awards for his work in the field of literature and the arts.

Use the following questions to assist you in your reading:

1. This reading makes the point that our economic system is partially responsible for our ecological problems. What exactly is this relationship as Mr. Krutch desribes it?
2. How does he use the example of gasoline taxes to illustrate the problem?
3. What solutions does he propose for the problem he raises?

*Joseph Wood Krutch, *If You Don't Mind My Saying So . . .*, **The American Scholar,** Summer, 1970, pp. 378-84. Reprinted from **The American Scholar,** Volume 39, Number 3, Summer, 1970.

Flipping through the pages of a current issue of a women's magazine, I found an advertisement for Chux, a disposable diaper. Its trade name is a recognition of the fact that "produce more and more" means also "get rid of more and more." The old formula for the good citizen, "produce more and consume more," has become "produce more and chuck more." . . .

What we are just beginning to realize is the false implication of the word disposable. Matter is indestructible, not disposable — or at least was until Einstein came along with his $e = mc^2$, and it will remain so for practical purposes until some method is discovered for applying the magic formula to a garbage dump — although if it ever is I don't know what we will do with all the energy generated. What we manufacture we are not even in practical terms disposing of. We are merely moving it somewhere else or turning it into something different but still not disposable — even when it goes up in smoke to come down again in smog.

No one has really faced the fact. We are still determined to increase every day the gross national product, forgetting that increasing the gross national product means increasing the gross national junk pile, since all our products will end up ultimately in the junk piles that surround our cities, poison the water we drink and foul the air we breathe.

The diaper, to take again this humble example, is flushed down the toilet and is for the moment out of sight. But it will surely turn up again, often in the untreated sewage that thousands of the smaller communities are pouring directly into the streams, clogging them with organic slime and poisoning the fish. Far from recognizing this fact, the same company that advertises the disposable diaper boasts on its television commercial that this is only the beginning. The baby's crib and ultimately all the furniture of his nursery will be disposable. You cut down trees to make more refuse and thus you keep the gross national product booming, whereas if you made something usefully durable, you would save the trees and relieve, instead of further overburden, the polluted environment. But that would cause prosperity to falter, so they say.

The solution, if there is one, would seem to mean one or all of the following: 1) a reduced population, which would therefore not need so much; 2) willingness to get along with less; 3) finding some way of making our products actually, for all practical purposes, disposable; 4) developing some system of recycling the residue.

Nature creates no junk piles. What she herself produces is not disposable but reusable. Whatever grows also decays. The complex becomes simple again, then is raised to complexity once more, and so the eternal recycling goes on.

It has become apparent that what is really needed today is a reevaluation of our life-styles. Our society has consumed inordinate amounts of material goods which have caused the environmental crisis we presently face. This gross consumption of energy and goods cannot continue if we are to live in a quality environment. We must become aware that the quality of our lives is more important than the quantity of material goods that we consume. We must learn to live with nature instead of trying to dominate her. Our values must change to take into account the necessity and beauty of trees and open spaces, and we must begin to find meaning in our lives in areas other than the number of our possessions.

From **Energy: Environmental Action Papers Number 1**, produced by a grant from Public Service Company of Colorado.

But a very large part of all that industry produces remains outside this recycling, and is neither reused nor, ultimately, disposable. Moreover, every advance in the efficiency of production along with every rise in the cost of human labor tends to diminish the extent to which anything is reused or recycled. Because it is cheaper to produce steel from newly mined ore than to reuse scrap, the abandoned automobiles threaten to engulf the cities, while the mines add to the pollution of the streams.

The returnable and reusable milk, soft drink or beer bottle is replaced by waxed "disposable" cartons or by metal containers which can be manufactured more cheaply than the returnable bottles can be handled and cleaned. Moreover, the so-called disposable tends to become more and more resistant to any recycling that nature can bring about — as has recently been pointed out is conspicuously the case where the tin can, which would disintegrate in ten or fifteen years, has been succeeded by the now-popular aluminum can, which can last for millenia. Ecology means literally housekeeping, but ours has been largely of the sweep-it-under-the-carpet kind, and it won't work.

The citizen in the industrialized country thinks he can look down upon the system of man, animals and subsistence agriculture that provides some living from an acre or two in India when the monsoon rains are favorable. Yet if fossil and nuclear fuels were cut off, we would have to recruit farmers from India and other underdeveloped countries to show the now-affluent citizens how to survive on the land while the population was being reduced a hundredfold to make it possible.

Howard Odum in **Environment, Power and Society**.

The fundamental fact is that you cannot solve the problems of pollution and environmental deterioration — or adopt an ecological rather than technological view — by merely giving them some thought while still accepting prevailing values. It will often happen that the only ecologically sound procedure is economically unsound. The two philosophies meet head on. They cannot be reconciled. We must make some sort of choice between them. To put it into even simpler terms, we would just have to stop asking, "Is this the cheapest way to make something or do something?" but ask instead, "Which way will leave the least residue of one kind or another, even if (as will usually be the case) this means also a loss of convenience?" . . .

If you suggest, for example, that gasoline taxes should be diverted from road building (which solves no problems) to the development of public transportation, you will surely be greeted by an agonized protest from the oil companies and the auto manufacturers who will tell you, not only that they ought to get the benefit of the taxes their customers pay, but that any significant decline in the auto industry would have catastrophic effects on the total economy — which I suppose is what it would do. Moreover, it is extremely doubtful that the public would submit to the inconvenience of the bus as opposed to the convenience of private transporation. In fact, even such public-minded citizens as you and I would object! What is to be expected of the less enlightened? Ask an economist if we can afford to do what would be necessary to reverse the current trends, and he will certainly answer "No." Ask the ecologist if we can afford not to do so, and he also will answer "No." In a sense, both are right.

"You want business in this town or don't you?"

Perhaps the time will come when we will have to abandon the whole assumption that it is the chief duty of the citizen to consume and throw away in order that prosperity will continue to increase. Perhaps we will have to return to the old philosophy that assumes that the good citizen is the one who preserves and continues to use what his grandfather had bought with the conviction that it would last for generations.

AMERICA IS A GROWING COUNTRY

by Fred Smith*

Fred Smith, the author, is a business executive who has worked for many years in the related fields of conservation and outdoor recreation, and is an associate of Laurance S. Rockefeller and a consultant to the Rockefeller family.

He was appointed by President Eisenhower to the Outdoor Recreation Resources Review Commission and by the Secretary of the Interior to the National Parks Advisory Board. He served as a consultant to President Nixon's Citizens' Advisory Committee on Environmental Quality, of which Laurance S. Rockefeller is Chairman, is Chairman of the Federal Power Commission's Task Force on the Environment, Chairman of the State Park Commission for the City of New York, a member of the United Nations Development (public) Corporation Board of Directors, and sits on several corporate Boards.

Bring the following questions to your reading:

1. Why is economic growth necessary?
2. Who are the extremists and why are they dangerous?
3. What does the typical industrialist spend his life doing?
4. How can we reconcile the need for economic growth with environmental needs?

*Additional copies of this article may be obtained from Harry Rand, Room 5600, 30 Rockefeller Plaza, New York, N.Y. 10020.

Our system has let us down, we are hearing; industry is wrong-headed, worthless, and endlessly damaging to society; and we need to shift gears. Do away with growth, curb technology, they say, or our world will self-destruct in two generations. . . .

It is widely held these days by environmentalists and by many frightened by environmentalists, that we've "got to get off this growth kick" — that's the new cliche — get off this growth kick if we are to survive. We are told we've got to put a stop to "progress." No more power plants, especially nuclear power plants. No SST. No increase in the Gross National Product. No more great technological advances. Curtail the use of energy. If industrial plants pollute, shut them down. If they might present a problem, don't build them. If this throws men out of work, let them find other jobs. Let single-minded officials set ultimate anti-pollution standards and demand instant adherence, whether or not there are yet practical ways to do it. Put a stop to the mining of coal, prevent the drilling of oil wells, curtail highway construction. Go to court and delay anything that purports to be essential to growth of any description. This, they say, is the only way to preserve the world we live in.

But at what cost? . . .

A substantial factor of growth and a continuing need for technological advance is built into our system, just as surely as a progressive increase in our population is inescapable. Being necessary does not make progress desirable — it only makes it unavoidable, and no amount of rhetoric can change this fact of life. . . .

All environmentalists are not unreasonable transcendentalists with incurable tunnel vision — nor is every industrialist a stereotype profit-monger with disdain for the public interest. It is the *extremists* who are dangerous: the environmentalists who demand instant cures, and the industrialists who won't budge.

I have had a considerable amount of first-hand experience with industry and industrialists during this environmental era, and by and large I would give them good marks. Most of them are as much concerned about pollution, for example, as the most adamant environmentalist. Most want to move as fast as technology and economic feasibility permit. But because their destinies are determined by consumers, they are more sensitive to the effects of cost than the affluent environmental enthusiast who says, "Do as I say, whatever it costs, and raise your prices if necessary. The people will pay!"

But will they? *Should* they have no choice? . . .

During the past few years the objective of continued economic growth has been attacked from two main directions. First, it has been argued that further economic growth is undesirable since it doesn't really make people any happier. The fact that people in the Western world now live longer; are much healthier and better fed; have more material comfort, education, leisure, opportunities to travel, and so on; have a generally wider range of the cultural possibilities and are less ravaged by such diseases as TB, diphtheria and polio, are of little importance beside the minor inconveniences and loss of privileges which are resented by a segment of society.

Wilfred Beckerman, *The Myth of Environmental Catastrophe*, **The National Review**, November 24, 1972.

The typical industrialist spends his life developing technologies and perfecting methods to bring down the cost of what he makes to the point where more and more people can afford to buy, and in buying contribute to the flow of funds that creates more jobs and makes higher pay levels possible. Whether anybody likes it or not, that is what our economy is all about. That is why the number of poverty-level people has shrunk from 350 per thousand population in Roosevelt's time to about 125 today. That is how poor people get bailed out. In the end, it is the only way. And they know it. They *will* be heard from; and when this happens, our politicians will start talking out of the other side of their mouths — some already are. They will run in the other direction — too fast and probably too far.

Is it already too late to establish ground rules that can save us from backlash?

How can we insure that we can retain the good that has already come from the recognition that we need more quality in our lives? Isn't there some way that extremists can be curbed and cooperation encouraged? Can't we demonstrate that the American tradition of progress, of moving ahead, of aspiring, is not really all that incompatible with the protection and enhancement of the environment? . . .

We can set up machinery — preferably at the state level — to investigate thoroughly all major environmental controversies, consider all the positions and claims, and then settle controversies out of court, with full power to enforce the decision. We can discourage resorting to the already overburdened courts to resolve problems that can be far more constructively resolved by negotiation. We can encourage the media and politicians to spend less time fanning the fires of controversy and more time assessing the economic, social, and long-range ramifications of important environmental questions. We can encourage opinion leaders and activists generally to develop the assets of our society, which are many, instead of embellishing and exploiting its weaknesses, which also are many. We can and certainly *should* oppose the very considerable efforts that are in vogue to discourage technological advances on the basis that there might be in them some hidden or latent danger. . . .

In short, we can work together intelligently, constructively, effectively, and systematically to solve the environmental problems that we all know must be solved; and we can do it without, in the process, fatally wounding the Goose that Lays the Golden Eggs. . . .

Industry, commerce, technology — these are our tools, as well as our problems — and proper use of them is the only possible means by which a Quality of Life can be built in a nation of 200 million people — and growing. We can be assured that the era of quantity at any environmental cost is past; that the era of reverence for the environment is here to stay.

ABILITY TO DISCRIMINATE

Usually difficult situations fail to present easy choices. Real life problems are too complex to permit simple choices between absolute right and wrong. The following exercise will test your ability to discriminate between degrees of truth and falsehood by completing the questionnaire. Circle the number on the continuum which most closely identifies your evaluation regarding each statement's degree of truth or falsehood.

1. Our prime national goals should be to reach a zero economic growth rate and a zero population growth rate.

+	5	4	3	2	1	0	1	2	3	4	5	—
	Completely True			Partially True			Partially False			Completely False		

2. To successfully combat national pollution we will have to change some of our basic values.

+	5	4	3	2	1	0	1	2	3	4	5	—
	Completely True			Partially True			Partially False			Completely False		

3. The Judeo-Christian ethic is partially responsible for the current ecology crisis.

+	5	4	3	2	1	0	1	2	3	4	5	—
	Completely True			Partially True			Partially False			Completely False		

4. Capitalism is the cause of most of our pollution problems.

+	5	4	3	2	1	0	1	2	3	4	5	—
	Completely True			Partially True			Partially False			Completely False		

5. Most environmentalists are promoting government control of human reproduction.

+	5	4	3	2	1	0	1	2	3	4	5	—
	Completely True			Partially True			Partially False			Completely False		

6. Many environmentalists use the ecology issue as a pretense to advance Marxist political schemes.

+	5	4	3	2	1	0	1	2	3	4	5	—
	Completely True			Partially True			Partially False			Completely False		

7. The supply of oxygen is not threatened by man's activity in any significant way.

+	5	4	3	2	1	0	1	2	3	4	5	—
	Completely True			Partially True			Partially False			Completely False		

8. Growth and technological advancement are natural and inevitable parts of our system.

+	5	4	3	2	1	0	1	2	3	4	5	—
	Completely True			Partially True			Partially False			Completely False		

THE POPULATION CONTROVERSY

"AH-HA! - NOW I SEE THE ENEMY!"

Readings

9. The Population Bomb
   Paul Ehrlich
10. The Nonsense Explosion
    Ben Wattenberg
11. A Better World for Fewer Children
    George Wald
12. Encyclical on Birth Control
    Paul VI

READING NUMBER 

THE POPULATION BOMB

by Paul Ehrlich*

Dr. Paul R. Ehrlich specializes in population biology. He is a Professor of biology and Director of Graduate Study for the Department of Biological Sciences, Stanford University. Dr. Ehrlich has written many papers and several books concerning the dangers of over-population and related matters.

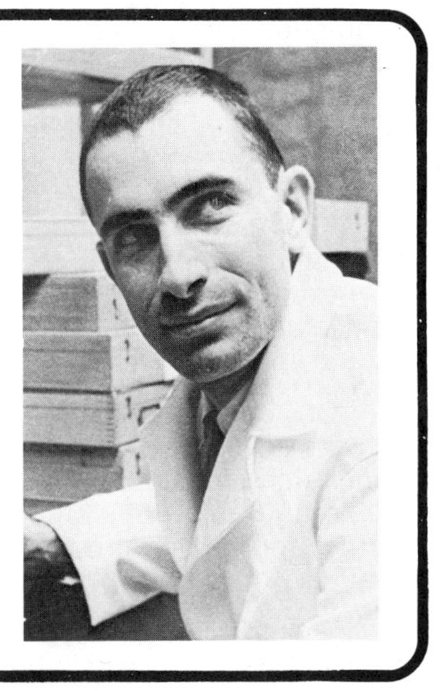

Consider the following questions while reading:

1. How does the author feel over-population will influence American society? Do you agree?
2. What proposals does he make?
3. Why does Dr. Ehrlich criticize the idea of an ever-expanding economy fueled by population growth?

*Dr. Paul R. Ehrlich, **The Population Bomb** (New York: Ballantine Books, Inc., 1968), pp. 132, 133, 135, 149, 150, 151, 152. Reprinted with permission from Ballantine Books, Inc.

The key to the whole business, in my opinion, is held by the United States. We are the most influential superpower; we are the richest nation in the world. At the same time we are also just one country on an ever-shrinking planet. It is obvious that we cannot exist unaffected by the fate of our fellows on the other end of the good ship Earth. If their end of the ship sinks, we shall at the very least have to put up with the spectacle of their drowning and listen to their screams. . . .

We, of course, cannot remain affluent and isolated. At the moment the United States uses well over half of all the raw materials consumed each year. Think of it. Less than 1/15th of the population of the world requires more than all the rest to maintain its inflated position. If present trends continue, in 20 years we will be much less than 1/15th of the population, and yet we may use some 80% of the resources consumed. Our affluence depends heavily on many different kinds of imports: ferroalloys (metals used to make various kinds of steel), tin, bauxite (aluminum ore), rubber, and so forth. Will other countries, many of them in the grip of starvation and anarchy, still happily supply these materials to a nation that cannot give them food? Even the technological optimists don't think we can free ourselves of the need for imports in the near future, so we're going to be up against it. But, then, at least our balance of payments should improve! . . .

Obviously our first step must be to immediately establish and advertise drastic policies designed to bring our own population size under control. We must define a goal of a stable optimum population size for the United States and display our determination to move rapidly toward that goal. . . .

The idea of an ever-expanding economy fueled by population growth seems tightly entrenched in the minds of businessmen, if not in the minds of economists. Each new baby is viewed as a consumer to stimulate an ever-growing economy. Each baby is, of course, potentially one of the unemployed, but a consumer nonetheless. The Rienows[43] estimate that each American baby will consume in a 70-year life span, directly or indirectly: 26 million gallons of water, 21 thousand gallons of gasoline, 10 thousand pounds of meat, 28 thousand pounds of milk and cream, $5,000 to $8,000 in school building materials, $6,300 worth of clothing, and $7,000 worth of furniture. It's not a baby, it's Superconsumer!

[43] **Moment in the Sun**, Dial Press, Inc., N.Y., p. 3.

The Commission on Population Growth and the American Future, after a two-year study by specialists in the various fields, stated that "we have looked for, and have not found, any convincing economic argument for continued population growth."

Furthermore, the Commission says, "a reduction in the rate of population growth would bring important benefits."

There is a great wealth of historical evidence to support these contentions. Japan, for example, has been for a decade the nation of greatest economic growth, at the very time when official policies were reducing the birth and growth rates to a level only a little higher than that of the United States and other industrialized countries.

European nations, on the whole, have very low growth rates and yet have been strong economically. There seems to be no connection in Europe between population growth and the economy, unless one would say that low population growth is good for the economy.

On the other hand, the nations with the greatest population growth are almost universally either static or declining economically.

Willard Johnson, Chairman, National Board, Zero Population Growth.

Our entire economy is geared to growing population and monumental waste. Buy land and hold it; the price is sure to go up. Why? Exploding population and finite resources. Buy automotive or airline stocks; their price is sure to go up. Why? More people to move around. Buy baby food stocks; their price is sure to go up. Why? You guess. And so it goes. Up goes the population and up goes that magical figure, the Gross National Product (GNP). And, as anyone who takes a close look at the glut, waste, pollution, and ugliness of America today can testify, it is well-named — as gross a product as one could wish for. We have assumed the role of the robber barons of all time. We have decided that we are the chosen people to steal all we can get of our planet's gradually stored and limited resources. To hell with future generations, and to hell with our fellow human beings today! We'll fly high now — hopefully they'll pay later. . . .

"AH-HA! - NOW I SEE THE ENEMY!"

Ways must be found to promote the idea that problems associated with population growth will more than cancel the "advantages" of financial prosperity. Perhaps the best way to do this would be to encourage Americans to ask exactly what our financial prosperity is for. What will be done with leisure time and money when all vacation spots are crowded beyond belief? Is it worth living in the Los Angeles smog for

50 weeks in order to spend two weeks in Yosemite Valley — when the Valley in the summer may be even more crowded than L.A. and twice as smoggy? What good is having the money for a fishing trip when fish are dead or poisonous because of pesticide pollution? Why own a fancy car in which to get asphyxiated in monster traffic jams? Do we want more and more of the same until we have destroyed ourselves? Sizable segments of our population, especially the young, are already answering that question: *"Hell, no!"* Their response should be considered carefully by population-promoting tycoons. . . .

Legal steps must be taken, and taken fast, to see to it that polluters pay through the nose for their destructive acts. The old idea that industry could create the mess and then the taxpapers must clean it up has to go. The garbage produced by an industry is the responsibility of that industry. The government should not use other people's money to clean it up. Keep the government out of business. Let it play its proper role in a capitalistic society — seeing to it that all segments of private enterprise do business honestly, seeing to it that the interests of the fishing industry are not subordinated to those of the petrochemical industry, seeing to it that your right to swim in a public lake is not subordinated to the desire of a steel company to make an inflated profit. . . .

READING NUMBER 10

THE NONSENSE EXPLOSION

by Ben Wattenberg*

Ben Wattenberg served on President Johnson's White House staff. He co-authored a book on demography titled **This U.S.A.**, published by Doubleday in 1965. He also co-authored **The Real Majority**, published in 1970.

Use the following questions to assist you in your reading:

1. Why does the author say that America is not a crowded country?
2. Why does he claim that more people will not create a crisis for our country?
3. How does Wattenberg feel the growth of population will affect our natural resources?

*Ben Wattenberg, *The Nonsense Explosion*, **The New Republic,** April 4 & 11, 1970, pp. 18-23. Reprinted by permission of **The New Republic,** © 1970, Harrison-Blaine of New Jersey, Inc.

Crowded, crowded, crowded, we are told. Slums are crowded, suburbs are crowded, megalopolis is crowded and more and more and more people are eating up, burning up and using up the beauty and wealth of America — turning the land into a polluted, depleted sprawl of scummy water and flickering neon, an ecological catastrophe stretching from the Everglades to the Pacific Northwest. Crisis. Crisis. Crisis.

That so very much of this is preposterous, as we shall see, should come as no real surprise to those who follow the fads of crisis in America. . . .

The critical facts are that America is not by any standard a crowded country and that the American birth rate has recently been at an all-time low.

The critical premise is that population growth in America is harmful.

In not stating the facts and in not at least challenging the premises, politicians and planners alike seem to be leaving themselves open to both bad planning and bad politics. This happens by concentrating on what the problem is not, rather than on what the problem is. Let's, then, first look at the facts. The current population of the United States is 205 million. That population is distributed over 3,615,123 square miles of land, for a density of about 55 persons per square mile. In terms of density, this makes the United States one of the most sparsely populated nations in the world. . . .

But while it is of interest to know that America has some land that is uninhabitable, what is of far more importance is that we have in the United States vast unused areas of eminently habitable land, land that in fact was inhabited until very recently. . . .

We can now turn to the premise set forth by the Explosionists i.e., more Americans are bad.

Are they? My own judgment is — not necessarily. . . .

The Explosionists say people, and the industry needed to support people, causes pollution. Ergo: fewer people — less pollution.

On the surface, a reasonable enough statement; certainly, population is one of the variables in the pollution problem. Yet, there is something else to be said. People not only cause pollution, but once you have a substantial number of people, it is only people that can solve pollution.

Further, the case can be made that, *more people* can more easily and more quickly solve pollution problems than can fewer people. For example: let us assume that $60 billion per year are necessary for national defense. The cost of defense will not necessarily be higher for a nation of three hundred million than for a nation of two hundred million. Yet the tax revenues to the government would be immensely higher, freeing vast sums of tax money to be used for the very expensive programs that are necessary for air, water, and pollution control. Spreading constant defense costs over a large population base provides proportionately greater amounts for nondefense spending. The same sort of equation can be used for the huge, one-time capital costs of research that must go into an effective, long-range anti-pollution program. The costs are roughly the same for 200 or 300 million people — but easier to pay by 300 million. . . .

Next, the Explosionists view more people as a crisis because of all the demands they will make upon the society. So many new schools, so many more hospitals, more libraries — services and facilities which we are having difficulty providing right now. Similarly with "new towns." If we are to avoid vast and sprawling megalopolitan swaths, we are told, we must build 100 brand-new towns in 30 years. Unfortunately, we've only been able to construct a few in the last couple of decades — so, alas, what possible chance do we have to make the grade in the years to come?

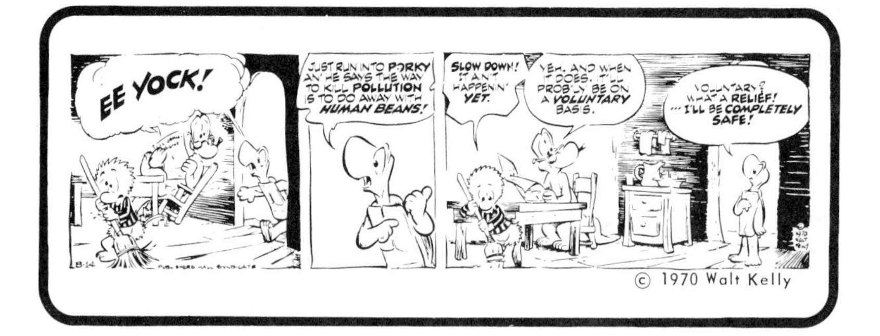

© 1970 Walt Kelly

What this argument ignores, of course, is that it is not governments who really create schools, hospitals, libraries and even new towns. It is *people* who create and build. People pay taxes; the taxes build and staff the schools; the more people, the more need for schools, *and* the more taxes. In an uncanny way it usually works out that every child in America has his own set of parents, and a school to attend. In a nation of a hundred million there were roughly enough schools for the children then present, at two hundred million the same was true and, no doubt, it will hold true at three hundred million. Nor will quality suffer because of numbers; quality suffers if taxpayers aren't willing to pay for quality and it is not harder for 300 million Americans to pay for quality schools for their children than it is for 230 million to buy quality schooling for their offspring. . . .

There is next the "resources" argument. It comes in two parts. Part one: many of our resources are finite (oil, coal, etc.); more people obviously use more resources; the fewer the people, the less the drain on the resources. Part two: we Americans are rich people; rich people use more resources; therefore, we must cut back population particularly fast, and particularly our rich population.

The resources problem is difficult to assess. A demographer now in his sixties seemed to put it in perspective. "*Resources are a serious problem,*" he said, "*We've been running out of oil ever since I was a boy.*"

The fact is, of course, sooner or later we *will* run out of oil; perhaps in thirty years or fifty years, or a hundred years or two hundred years. So too will we run out of *all* non-renewable resources — by definition. We will run out of oil even if population growth stops today and we will run out of oil, somewhat sooner, if population growth continues. Whether oil reserves are depleted in 2020 or 2040 or 2140 does not seem to be of critical importance; in any event a substitute fuel must be found — probably nuclear. If no adequate substitute is developed, then we (all of us earthmen) will suffer somewhat regardless of numbers.

Part two, that *rich* people are the real menace both resource-wise and pollution-wise, has recently been particularly stressed by Dr. Jean Mayer who advises the President hunger-wise but would not seem to be fully up to date demography-wise.

For the simple fact is that wealthier people generally have far fewer children than poorer people. With current mortality rates, population stability is maintained if the typical woman has on the average 2.13 children. In a 1964 Census Bureau survey among women who had completed their child-bearing years, it was shown that families with incomes of $10,000 and over had 2.21 children, just a trifle over replacement. This compared with 3.53 children for the poorest women. Since 1964, fertility rates have gone down among young women, and it is possible that when these lower rates are ultimately reflected as "completed fertility" we may see that affluent American women of the future just barely replace their own number, if that.

In short, current population patterns show that affluent people do not cause rapid population growth. And if the entire population were entirely affluent, we certainly would not be talking about a population explosion. Further, if the entire population were affluent *and* committed to combatting pollution, we wouldn't be talking about a pollution explosion either.

America could support twice its current population and probably four times its current population.

What then is Dr. Mayer's prescription? Is he against affluent people having babies but not poor people, even though the affluent have relatively few anyway? Or perhaps it is that he is just against the idea of letting any more poor people become affluent people, because they too will then consume too many resources and cause more pollution?

There are two important points that run through most of the above. First is that the simple numbers of people are not in themselves of great importance in the United States. There is no "optimum" population as such for the U.S., not within population ranges now forecast in any event. Whether we have 250 million people or 350 million people is less important than what the people — however many of them there are — decide to do about their problems. Second the population problem, at least in the United States, is an extremely long-term proposition, and in a country of this size and wealth, there is more flexibility in solving the potential demographic problems than might be assumed from the current rhetoric-of-crisis. . . .

Certainly too, population growth must sooner or later level off. While America could support twice its current population and probably four times its current population — growth can obviously not go on forever and it is wise to understand this fact now rather than a hundred years from now. . . .

But what is wrong, and dangerous, and foolhardy is to make population a crisis. Doing so will simply allow too many politicians to take their eyes off the ball. When Explosionists say, as they do, that crime, riots, and urban problems are caused by "the population explosion," it is just too easy for politicians to agree and say sure, let's stop having so many babies, instead of saying let's get to work on the real urban problems of this nation. (As a matter of general interest it should be noted that the riot areas, the high-crime areas, the areas of the most acute urban problems are *areas that are typically losing population.* For example, special censuses in Hough and Watts showed population *loss*. Given that kind of data it is hard to accept the Explosionist notion that crowding causes crime.) . . .

When the Explosionists say, as they do, that it's because we have so many people that Lake Erie is polluted then once again we are invited to take our eye off the tens-of-*billions*-of-dollars ball of environmental safety and we are simultaneously invited to piddle around with 25-*million* dollar programs for birth control, which are nice, but don't solve anything to do with Lake Erie.

Finally, we must take note of the new thrust by the Explosionists: population control. Note the phrase carefully. This is specifically not "family planning," where the family concerned does the planning. This is *control* of population by the government and this is what the apocalyptics are demanding, because, they say, family planning by itself will not reduce us to a zero growth rate. The more popular "soft" position of government control involves what is called "disincentives," that is, a few minor measures like changing the taxation system, the school system and the moral code to see if that won't work before going onto outright baby licensing. . . .

What it all adds up to is this: why have a long-range manageable population problem that can be coped with gradually over generations when, with a little extra souped-up scare rhetoric, we can drum up a full-fledged crisis? We certainly need one; it's been months since we've had a crisis. After all, Vietnam, we were told, was "the greatest crisis in a hundred years." Piker. Here's a crisis that's a beauty: the greatest crisis in two billion years; we're about to breed ourselves right into oblivion. . . .

EXERCISE 5

FACT AND OPINION

This discussion exercise is designed to promote experimentation with one's ability to distinguish between fact and opinion. It is a fact, for example, that the United States was militarily involved in the Vietnam War. But to say this involvement served the interests of world peace is an opinion or conclusion. Future historians will agree that American soldiers fought in Vietnam, but their interpretations about the causes and consequences of the war will probably vary greatly.

Some of the following statements are taken from the two preceding readings and some have other origins. Consider each statement carefully. Mark (O) for any statement you feel is an opinion or interpretation of the facts. Mark (F) for any statement you believe is a fact. Then discuss and compare your judgments with those of other class members.

O = Opinion
F = Fact

___ 1. There is a population crisis in the United States today.

___ 2. The United States consumes more of the world's natural resources than any other country.

___ 3. Our economy is geared to a growing population and monumental waste.

___ 4. The United States has vast unused areas of habitable land.

___ 5. Rich people cause more pollution than poor people.

___ 6. We have a long-range manageable population problem that can be dealt with gradually over generations.

___ 7. The pollution produced by private industry is the responsibility of industry. The government should not use the taxpayers' money to clean it up.

___ 8. Continuous economic growth and expansion will produce a continuous growth in social well-being.

___ 9. The world population growth rate could lead to the future extinction of the human race.

___ 10. The federal government has the primary responsibility to solve our nation's environmental problems.

___ 11. The United States is the world's major polluter.

CITIZEN ON A TIGHTROPE

READING NUMBER 11

A BETTER WORLD FOR FEWER CHILDREN

by George Wald*

George Wald, Harvard biology professor and Nobel Laureate in physiology and medicine became nationally known following a talk he gave at Massachusetts Institute of Technology dealing with the influence of the military on American society. He is currently Higgins professor of biology.

The following questions should help you examine the reading:

1. What point is Wald attempting to make in his statement "a better world for fewer children"?
2. What specific measures does Wald suggest to limit population growth?
3. What complications does Wald foresee in suggesting abortion as a means of limiting population growth?

*George Wald, *A Better World for Fewer Children*, **The Progressive,** April, 1970, pp. 26-28. Reprinted with permission from George Wald.

None of the things that now most need to be done for our country and for the world have much chance of working unless coupled with the control of population. By present indications our present population of 3.5 billions will have doubled by the end of this century. Long before that, we can expect famine on an unprecedented scale in many parts of the world.

Yet this in itself is not the heart of the problem. If it were, one would have some small reason for optimism; for in the last decade the world's food supplies have increased more rapidly than its population.

The concept that food is the primary problem is a prevalent and dangerous misunderstanding. It implies that the main point of the human enterprise from now on will be to see how many persons can be kept alive on the surface of the earth. A distinguished demographer recently estimated that with what he calls "proper management" we could support a world population of forty billions.

Of course, under those conditions people would not eat meat; there would be no place for cows, sheep, or pigs in such a world. This would be an altogether bankrupt view of the human enterprise. Humanity still has a chance at creating an ever wider, richer, and more meaningful culture. This would degrade it to simple production — a meaningless venture in simple multiplication. Even that, however well managed, must come to an end, as the potential resources of the planet become insufficient to feed further numbers of people.

The point then is not how much people one can feed on this planet, but what population can best fulfill human potentialities. One is interested not in the quantity but in the quality of human life. From that point of view the world is probably already over-populated. China and India were once great cultures, enormously creative in the sciences, the visual arts, and literature; but those aspects of Indian and Chinese culture declined centuries ago for reasons associated, I think, with overpopulation.

The Western world also is becoming crowded; I do not think it irrelevant that the quality of our production in the arts has declined greatly in the last century or two. Western science is flourishing, but the productivity of the individual scientist is nothing like what it was up to a century ago. It would take four to six top scientists of the present generation to approach in productivity, scope, and quality the contributions of Charles Darwin, Hermann von Helmholtz, or

James Clerk Maxwell. Those men had no labor-saving devices, so far as I know not even secretaries: no dictaphones, microfilm, computerized information, retrieval services, and the like. They did, however, have peace and quiet, the chance to walk through green fields, along quiet rivers, and to find relief from all the crowding, noise, filth, and endless distractions of modern urban life.

The earth is a small planet rapidly running out of food and energy. World population grows two percent annually, doubling every 35 years. There are 3.5 billion people on Spaceship Earth now and there will be twice that by A.D. 2000. Take a day off; when you come back the earth will have 200,000 more people. That's the size of Des Moines. We add a new Des Moines to our spaceship's passenger list every 24 hours. This can't last. It's as though our pleasant planet were infected — infected with people. If we don't curb our procreation voluntarily Nature will do it some other way, by war or pestilence or famine. Probably famine.

From the TRB column, **The New Republic**, June 16, 1973.

A second profound misunderstanding has plagued many earlier discussions of the population problem. This is the widely prevalent view that the poor are over-reproducing, the well-to-do are under-reproducing, and the quality of the human race is hence going downhill. Quite apart from the naive assumption that the economically poor are necessarily also genetically inferior, there are other, almost as serious trouble with this view of the problem.

We are beginning to realize now that it is precisely the well-to-do and their children who make the most trouble — who are at once the biggest consumers and the biggest polluters. That is true individually, and has its national aspects. It is claimed by reliable sources that an American child uses fifty times as much of the world's resources as an Indian child. Our country, which contains only about six per cent of the world's population, uses about forty per cent of the world's resources, and accounts for about fifty per cent of the world's industrial pollution.

So it is essential that we bring world population under control, not only to keep it from increasing further, but if possible cut it down from its present level. That won't be easy or altogether pleasant. Indeed, it will be so difficult that we would be well advised to choose any viable alternative. But there is none. It is that or disaster. We are not being asked, but told to control world population. That is now our only chance of a meaningful survival. And whatever is done now must be done quickly, not only because the population is increasing so explosively, but because as one consequence the quality of human life has already been eroded. We must be aware of the danger that persons of future generations, even more out of contact than we with the potentialities of a less crowded world, will have lost a wider human view, and will have become unable to help themselves. In that sense we may be the last generation that can save humanity.

So what to do? First, as rapidly as possible make convenient, safe, and cheap — I would rather say free — means of birth control universally available. That, however, will certainly not be enough. I think we must as rapidly as possible make convenient, safe, altogether legal, and cheap — again I would rather say free — means of abortion universally available.

I hope for the early advent of a safe and efficient abortion pill. There are recent reports of encouraging work in this direction, from England and Sweden, with one of the prostaglandins. I think the condition we must try to achieve everywhere and as rapidly as possible is to see to it that nowhere in the world need a woman have an unwanted child. Having got there, we can take stock and see whether that is yet enough. Very likely not; and then we shall have to go on with a variety of other reasonable procedures. We might begin some of those procedures much sooner: for example, legislate tax discouragements rather than incentives for bearing children, particularly beyond the first two.

Many people still have trouble with the thought of legalized abortion. Of course the Roman Catholic Church is deeply opposed to it. Lately it has at least considered accepting contraception, before officially deciding against it. It seems to me likely that having once opened the question of contraception, it will now prove very difficult to close.

Abortion, however, is not only rejected by the Catholic hierarchy, but apparently also by large numbers of Catholic laymen. They regard abortion as highly immoral, indeed a form of murder.

It is difficult for me to appreciate the morality of that position in view of the present condition of the world's children. If we were in fact taking proper care of children all over the world, raising them with enough food and shelter and clothing so that they had the chance to fulfill their genetic potentialities, then we might have the privilege of feeling that every embryo should be born. As it is, however, the world's population is now mainly held down through infant mortality. What is killing those children is war, famine, disease, and poverty. That is our present condition, one in which we turn the Four Horsemen of the Apocalypse loose upon the children of the earth. Surely we can and must do better than that.

It is not as though we were asked to introduce abortion into a world that is not already practicing it. It is in fact practiced very widely, not only in those few nations where it is legal, but in others where it is illegal, and frowned upon by tradition and religion. Indeed, some recent statistics seem to show that it is particularly prevalent in a number of Roman Catholic countries, in which other means of birth control are not available.

Data on the extent of illegal abortion are difficult to obtain and not altogether reliable. However, a recent study estimates that in France there is one abortion for each live birth; that in Latin America as a whole there is one abortion for every two live births; and that in Uruguay, there are three abortions to each live birth. (Alice S. Rossi, U.S. Public Health Service, writing in *Dissent*, July-August, 1969).

In numbers of underdeveloped countries, through ignorance, poverty, and the low state of technology, abortion is the principal method of birth control. A lot of this happens under brutal circumstances. The women do it to themselves, or to one another, or at best with the help of some self-taught midwife. All of them suffer, and many of them die. It is one of the penalties of poverty. What goes as morality, in this as in so many other things, sits more lightly on the well-to-do. The poor must bear it in suffering and terror, and at times must pay for it with their lives.

It is precisely a high concern for human life, and most of all for children and what becomes of them, that makes me believe that we must achieve as rapidly as possible universally available, and preferably free, birth control and abortion. Being born unwanted is no favor to any child. Being born to hunger, want, disease, and the ravages of total war is no favor to any child. We need to make a world in which fewer children are born, and in which we take better care of all of them.

So that is my program: *a better world for fewer children.*

READING NUMBER 12

ENCYCLICAL ON BIRTH CONTROL

by Paul VI*

Following are excerpts from the official English translation of Pope Paul's encyclical on birth control, Humanae Vitae.

As you read try to answer the following questions:

1. The previous reading advocates artificial methods of birth control to limit population growth. Why does the Pope oppose this suggestion?
2. What is the Pope's position on abortions for therapeutic reasons?

Fear is shown by many that world population is growing more rapidly than the available resources, with growing distress to many families and developing countries, so that the temptation for authorities to counter this danger with radical measures is great.

Moreover, working and lodging conditions, as well as increased exigencies both in the economic field and in that of education, often make the proper education of an elevated number of children difficult today.

Text of Encyclical on Birth Control*, **The Minneapolis Tribune, July 30, 1968. Reprinted, with permission, from **The Minneapolis Tribune.**

A change is also seen both in the manner of considering the person of woman and her place in society, and in the value to be attributed to conjugal love in marriage, and also in the appreciation to be made of the meaning of conjugal acts in relation to that love.

Finally and above all, man has made stupendous progress in the domination and rational organization of the forces of nature, such that he tends to extend this domination to his own total being: to the body, to physical life, to social life and even to the laws which regulate the transmission of life.

Broad Perspective Needed

The problem of birth, like every other problem regarding human life, is to be considered beyond partial perspectives — whether of the biological or psychological, demographic or sociological orders — in the light of an integral vision of man and of his vocation, not only his natural and earthly, but also his supernatural and eternal vocation.

And since, in the attempt to justify artificial methods of birth control, many have appealed to the demands both of conjugal love and of "responsible parenthood," it is good to state very precisely the true concept of these two great realities of married life, referring principally to what was recently set forth in this regard, and in a highly authoritative form, by the second Vatican Council in its pastoral constitution.

These acts, by which husband and wife are united in chaste intimacy, and by means of which human life is transmitted, are, as the council recalled, "noble and worthy" and they do not cease to be lawful if, for causes independent of the will of husband and wife, they are foreseen to be infecund, since they always remain ordained towards expressing and consolidating their union.

In fact, as experience bears witness, not every conjugal act is followed by a new life. God has wisely disposed natural laws and rhythms of fecundity which, of themselves, cause a separation in the succession of births.

Nonetheless the church, calling men back to the observance of the norms of the natural law, as interpreted by her constant doctrine, teaches that *each and every marriage act must remain open to the transmission of life.*

As experience bears witness, not every conjugal act is followed by a new life. God has wisely disposed natural laws and rhythms of fecundity which, of themselves, cause a separation in the succession of births.

Design of Marriage

It is in fact justly observed that a conjugal act imposed upon one's partner without regard for his or her condition and lawful desires is not a true act of love, and therefore denies an exigency of right moral order in the relationship between husband and wife.

Hence, one who reflects well must also recognize that a reciprocal act of love, which jeopardizes the responsibility to transmit life which God the Creator, according to particular laws, inserted therein, is in contradiction with the design constitutive of marriage, and with the will of the Author of Life.

To use this divine gift destroying, even if only partially, its meaning and its purpose is to contradict the nature both of man and woman and of their most intimate relationship, and therefore it is to contradict also the plan of God and His will.

In conformity with these landmarks in the human and Christian vision of marriage, we must once again declare that the direct interruption of the generative process already begun, and above all, directly willed and procured abortion, even if for therapeutic reasons, are to be absolutely excluded as licit means of regulating birth.

Equally to be excluded, as the teaching authority of the church has frequently declared, is direct sterilization, whether perpetual or temporary, whether of the man or of the woman.

Similarly excluded is every action which, either in anticipation of the conjugal act, or in its accomplishment, or in the development of its natural consequences, proposes, whether as an end or as a means, to render procreation impossible.

ABILITY TO EMPATHIZE

The ability to empathize, to see a problem from another person's vantage point, is a skill that must be widely developed and practiced if national problems, like conflict over abortion laws, are ever to be solved.

Consider the following proposals by Senator Robert W. Packwood of Oregon:

> *My proposal is strictly voluntary. If a woman chooses not to use a contraceptive device, that is her business. But the government has a responsibility to make birth control information available to all women. If the woman then chooses to use a device to prevent pregnancy, the government should make certain that she has access to such a device. If she cannot afford to pay for it, the government should provide it free of charge.*
>
> *The second voluntary step is to allow any woman in any state to have an abortion if she chooses.*
>
> *A third step which must be implemented is a government policy of encouraging smaller families through a tax incentive program. One way this can be done is by limiting the number of children which can be declared as income tax deductions.* *

Try to imagine how the following individuals might react to each of Senator Packwood's three proposals.

A liberal Roman Catholic wife

An atheist

A Lutheran minister

A congressman who represents a large rural area

A congressman who represents a large urban area

A presidential candidate

The president of Gerber Products (baby foods)

The president of a toy manufacturing company

A businessman in Sweden

THE ENERGY CRISIS

Readings

READING NUMBER **13**

VIEWS OF AN ENVIRONMENTALIST

by Michael McCloskey*

As executive director of the Sierra Club, Michael McCloskey delivered the following speech to the American Nuclear Society in Boston, Massachusetts, June 15, 1971.

Consider the following questions while reading:

1. Why does the author claim present rates of energy growth are unrealistic?
2. Why is nuclear power not the answer to our energy needs?
3. What suggestions does the author make to solve the problem of energy growth?

*Michael McCloskey, *The Energy Crisis*, **Vital Speeches**, August 1, 1971, pp. 621-24. Reprinted with permission from **Vital Speeches**.

In discussing any public issue, one is always confronted with the problem of defining what the issue is. What is the nature of the issue embodied in the "energy crisis"? The industries supplying energy would have us believe that the problem is one of expanding supplies and reducing constraints on growth. As an environmentalist on the other hand, I submit the problem is one of excessive pressures for growth in consuming energy.

I submit that present rates of energy growth are unrealistic, that they are environmentally damaging, and that they are artificially induced. I submit further that these rates cannot long continue, that our main task ought to be to bring these rates of growth under control, and that there are reasonable ways of doing this. . . .

Today we consume fifteen times the energy we did one hundred years ago, though our population has only tripled in that time. Over the past decade the average growth rate in the consumption of energy in all its forms has been more than 4 per cent annually, climbing to about 5 per cent annually over the last five years. Growth has been particularly phenomenal in the electrical energy sector, at about 7 per cent annually in recent years. Projections based on that rate of growth call for a doubling of electric power production about every ten years.

With these growth rates we may soon find that we are running out of space for power plants, to cite just one example. It has been calculated that even with large 1,000 megawatt power plants each requiring an area of only 1,000 feet on a side, in less than twenty doublings — less than 200 years — *all* the available land space in the United States would be occupied by such plants! In California where power production is expected to double every eight years, if power were to be supplied by 110 megawatt plants on 80-acre sites, the entire land area would be covered in only 122 years. Similar startling projections could doubtlessly be made for other forms of energy use; e.g., at projected rates of growth, how much space will need to be paved over to accommodate our automobile-oriented transportation system by various dates?

Other limits to proliferating energy use can be cited. For example, by the end of the century, if growth continues as projected at current rates, one-third of our total freshwater runoff might be required for power plant cooling purposes if only once-through cooling is used. If once-through cooling is superseded by cooling ponds and towers, then even more land is needed and the space crunch will come sooner.

WILL WE GO TO WAR?

a news release:

Americans soon may be confronted with the choice of accepting strict conservation measures on domestic oil production or sending an army to the Middle East to secure oil fields there, a federal official has warned.

Elmer F. Bennett, assistant director of the Office of Emergency Preparedness, told an energy symposium . . . that he hopes that the U.S. public will voluntarily adopt lower-horsepower engines and restraints on the use of other petroleum products, but he added:

"There are real doubts whether the American people are going to accept such changes in their lifestyle. . . .

"If our dependency on foreign oil becomes so great, or our control in foreign policy and international influence declines, we might someday see these measures adopted.

"The alternative would be to send an expeditionary force to the Middle East that would make Vietnam seem like a picnic."

Future War-Peace Oil Choice Warned, **Minneapolis Star**, April 12, 1973.

Ultimate limits to growth in energy use also obviously exist in the finite nature of our fuel resources. The fossil fuels now provide by far the greatest part of our energy sources (e.g. almost 96 per cent in 1969). Whatever the true situation as to immediate supplies, it is obvious that ultimately these nonrenewable resources will be depleted. Optimistic estimates predict that our fossil fuels as a group will be exhausted within a few hundred years at best, possibly much sooner. A recent National Academy of Sciences report, for example, predicts that within another fifty years or so the great bulk of the world's initial supply of recoverable petroleum liquids and natural gas will be exhausted, while recoverable fuel from the oil shales and tar sands might extend the lifetime of the petroleum group to a century or so. With respect to coal, the report estimates that if used as the principal source of energy at projected demands, it will last no more than two or three centuries.

Though nuclear power is expected to play a major role in future electrical energy production, electrical energy is of course only a part of the total energy consumed (presently, about one-quarter). And the supply of uranium 235 from high-grade ores is limited. The NAS report indicates that the production of nuclear power using the present type of reactors and uranium 235 as the principal energy source can be sustained for only a few decades. Another estimate gives high grade uranium ores a lifetime of under fifty years. Breeder reactors could extend these fuels, but it is not clear what the costs may be and operational and environmental problems are unsolved. And a practical method of producing electricity from fusion is still only a possibility. . . .

A short-run strategy would involve the following types of changes in public policy: ending or reducing the many biases in public policies which provide incentives to energy growth; maintaining and strengthening environmental constraints on energy growth; reducing energy demands by educating the public to understand the importance of conservative use of energy; encouraging intensified research and development on ways of achieving greater efficiencies in energy utilization and in finding new, more environmentally acceptable energy sources; discouraging growth in industries that are the most profligate consumers of energy; and establishing new agencies in government to plan and regulate activities. Each of these changes involves efforts that go well beyond the traditional bounds of energy policy, and all can have profound economic and social impacts. Yet changes are already beginning to occur in all these fields, and the environmental movement is determined to promote them. While each needs detailed elaboration, I will simply try to sketch some of the central ideas here.

(1) We will never be able to cope with the crisis of energy growth until we root out the many incentives in public policy to such growth. What would happen if the diseconomies of all these subsidies were to be removed or reduced? Quite likely the actual impetus behind energy

growth would drop sharply. We would no longer be artificially stimulating a false demand as if energy growth were a clear public good rather than a clear public problem.

(2) At the same time as artificial incentives are removed, positive constraints need to be asserted to protect environmental values. These include restrictions of various sorts on the modes of developing, processing, and transporting fuels: establishment of onshore and offshore closures to drilling and mining because of overriding environmental values, as in areas for wilderness, parks, refuges, marine sanctuaries, and for future study; strict environmental operating controls on open areas; a ban on strip mining; deferred development of oil shale; stringent controls on marine oil shipping, including controls on the operation and design of super tankers; stricter controls on pipeline specifications and routing; and tighter controls on the handling, processing, and disposal of nuclear waste products. Simultaneously, environmental controls need to be extended further on the facilities which consume fuels: principally power plants, automobiles, and basic industries. Pollution controls on effluents and emissions need to continue to be tightened, with greater emphasis on toxic substance and by-product recovery and closed-plant recycling. With the move toward national land-use controls, the siting of plants, transmission lines, and highways will become increasingly constrained.

The rising costs associated with all of these constraints can be regarded as an internalizing of social costs. As costs rise and are passed on to consumers, demand should slacken and the rate of growth in the drain on energy resources should slacken too.

(3) As the public faces higher energy and product costs, a strong effort needs to be made to promote public understanding of the reasons behind these shifts. As consumers, we all need to begin to understand that we have not been paying the full costs of driving automobiles and of using electricity almost as if it were a free good. Instead of looking upon these rising costs as a new burden upon the consumer we should look upon them as ending a "free ride" which we have been getting at the expense of the future and of the environment. Moreover, we should encourage even greater understanding of the need to end our wasteful habits in using energy. Changing cultural attitudes toward walking, bicycling, and mass transit can help break the omnipotence of the automobile as a symbol of fashion and prestige. Changing cultural attitudes can also help bring rising residential electrical use under control among the middle class; hopefully there will be a growing trend away from all-electric gadgetry as a status symbol.

Finally, however, it is important that steps be taken to make sure that rising prices do not block the aspirations of the poor. Particularly with respect to electrical rates, special steps should be taken to provide low rates for small residential consumers. Improved mass transit should also help low income urban residents. Other steps may need to be taken too to make sure that the poor are not penalized by these changes in public policy.

READING NUMBER **14**

ENVIRONMENTALISTS CONTRIBUTE TO ENERGY CRISIS

by M. Stanton Evans*

M. Stanton Evans is the editor of **The Indianapolis News**. He is a former assistant editor of **The Freeman** and managing editor of **Human Events Newsletter**, and is currently an associate editor and regular columnist for **The National Review**. His books include **The Liberal Establishment**, **Revolt on the Campus**, **The Politics of Surrender**, **The Lawbreakers**, and **The Future of Conservatism**.

Use the following questions to assist you in your reading:

1. What crisis have environmentalists produced?
2. What sources of energy are abundantly available in the U.S.?
3. What do ecologists want?

*M. Stanton Evans, *Environmentalists Contribute to Energy Crisis*, **Human Events**, May 26, 1973, p. 434. Reprinted with permission from **Human Events**.

Act in haste, repent at leisure is a familiar maxim of human folly with particular application to the world of politics.

Latest and most vivid case in point is the wave of manufactured hysteria about pollution which has swept the nation in recent years and produced a series of government decrees of such surpassing absurdity as to boggle the imagination. In the name of combating an illusory danger, the environmentalists and their friends in government have managed to concoct a real one.

That real disaster is the burgeoning crisis in matters of energy procurement which has by slow degrees been making its way into the headlines as Americans discover that gas station operators, providers of heating fuel and power utilities are curtailing supplies or expect to do so in the immediate future. The signs are becoming commonplace: Electrical brownouts, "save a Watt" campaigns, rationing of gasoline by major suppliers, shutdowns by independents, "voluntary guidelines" for conserving gasoline advanced by the federal planners, and so on.

> **For the leading industrial nation in the history of the world, this is a preposterous state of affairs — but an entirely predictable one. The environmentalists and the government regulators have been doing everything within their power to bring the wheels of industrial production to a halt, and they are perilously close to getting the job accomplished.**

It is true that fossil fuels on which the U.S. is now dependent for most of its energy are finite and will eventually be used up. But there are enough of them in combination to last us a couple of hundred years (i.e., for as long as the American Republic has existed to date), and in the meantime we have the capacity to develop new sources — principally nuclear power — which could supply our needs indefinitely. But the interference of the environmentalists has been steadily closing off our access to the old supply and simultaneously preventing development of the new.

An inventory of the power resources employed to fuel America's industrial machine reveals that 43 per cent of our energy is derived from oil, 32 per cent from natural gas, 21 per cent from coal, 4 per cent from hydroelectric power, and 1 per cent from nuclear power plants. Of these resources our coal supply alone is sufficient to power our economy for several hundred years or so depending on the uses to which it is put, gas and oil and coal together are certainly good for a couple of centuries, and the uses of nuclear power have hardly been tapped at all.

On that conspectus, there need not be an "energy crisis," but the environmentalists have willed it otherwise. With surgical thoroughness they have moved to choke off access to each of the major supplies, and in just about every case succeeded.

Clearly, we are facing a vitally important energy challenge. If present trends continue unchecked, we could face a genuine energy crisis. But that crisis can and should be averted, for we have the capacity and the resources to meet our energy needs if only we take the proper steps — and take them now.

More than half the world's total reserves of coal are located within the United States. This resource alone would be enough to provide for our energy needs for well over a century. We have potential resources of billions of barrels of recoverable oil, similar quantities of shale oil and more than 2,000 trillion cubic feet of natural gas. Properly managed, and with more attention on the part of consumers to the conservation of energy, these supplies can last for as long as our economy depends on conventional fuels.

In addition to natural fuels, we can draw upon hydroelectric plants and increasing numbers of nuclear powered facilities. Moreover, long before our present energy sources are exhausted, America's vast capabilities in research and development can provide us with new, clean and virtually unlimited sources of power.

Thus we should not be misled into pessimistic predictions of an energy disaster. But neither should we be lulled into a false sense of security. We must examine our circumstances realistically, carefully weigh the alternatives — and then move forward decisively.

From President Richard M. Nixon's message on energy policy, transmitted to the Congress on April 18, 1973.

One obvious energy source, for example, is offshore oil, but the ecologists have managed to hinder progress on this one by crying up an oil spill off Santa Barbara, Calif., causing cancellation of a number of existing leases and holding up the issuance of new ones. It is estimated that California's offshore reserves could supply the nation with two million barrels a day if the drilling moratorium there were lifted.

Ditto for Alaskan oil, which could bring us some two to three million barrels daily if we could overcome the eco-freak desire to preserve the "tundra" from a harmless pipeline. As Lawrence Rocks and Richard Runyon point out in their recent book, **The Energy Crisis** (Crown; $2.95), an oil spill from such a pipeline (an unlikely event) would damage approximately .01 per cent of the "tundra" and Alaskan forest. The dictionary definition of "tundra," incidentally, is a "treeless plain" which "consists of black mucky soil with a permanently frozen subsoil, but supports a dense growth of moss and lichens" (a lichen is a combination fungus and alga). Obviously, a priceless resource to be preserved though the heavens fall.

Not content to close down new sources of oil, the ecologists have also campaigned to use up the existing supply more rapidly — chiefly through the Clean Air Act of 1970, a masterpiece of regulatory fuddlement. Under its impossible standards, the auto industry has produced a breed of gas-consuming monsters precisely at the time that oil companies are feeling the pinch of shortage, while restrictions on power plants have caused them to turn increasingly from coal to cleaner-burning fuel oil. The ecologists have thus contrived to increase demand while ingeniously diminishing supply.

Coal is available in abundance, and can be obtained quite cheaply by surface mining. It can be burned directly or converted into synthetic oil and gas or used for the generation of electricity. Depending on its mix of uses, coal alone could supply our energy needs for 200 to 700 years. So naturally the friends of earth have moved to stop its production, and the government regulators have cooperated fully, cracking down on "strip mining" and refusing to lease government lands for mining even low-sulfur coal.

Electrical power is of course quite cheap and up until a couple of years ago increasingly plentiful, but the ecologists have got their hooks into that one, too. They have been industriously filing actions to stop construction of power plants under the National Environmental Protection Act —

and thus succeeded in hindering access to yet another source of power.

Consolidated Edison has been hung up for a decade by environmentalists in an effort to build a storage plant to improve electric service in New York — one example among many that could be cited. Look for bigger and better brown-outs in the future.

Another possible source of energy is nuclear power, but needless to remark the very mention of this (quite safe) expedient sends the environmentalists and the anti-nuclear fraternity into paroxysms of fear. Although the safety record of the civilian reactors already on stream is exemplary, the nation's cultivated campaign of terror against anything "atomic" has brought a flurry of suits against development of nuclear energy and slowed this program to a veritable crawl.

All of these stoppages have occurred with the regulatory complicity of government, and still other government actions have contributed to the problem: Regulation of natural gas prices at the well-head, encouraging consumption and limiting supply; import quotas on oil; reductions in the oil depletion allowances; punitive regulation of the railroads which are at once the most energy-efficient of common carriers and needed agents for the transportation of coal.

In all its dimensions, the energy crisis is the result of government interference — a fact which bodes exceeding ill for current proposals that government get more deeply involved in energy matters to bring the crisis to solution.

The ironic fact unstated in all this is that American industry over the past two decades has been producing more and better energy on every conceivable front, and has done so with a technology that has made the environment progressively cleaner instead of dirtier. That doesn't satisfy the ecologists, however; they want not simply a cleaner environment, but an absolutely clean one, which necessarily means an absence of energy and technological activity. A *totally* undisturbed environment, after all, would imply a world devoid of life and movement altogether — an outcome which begins to seem increasingly plausible.

Observing this performance, one is tempted to suspicions of metaphysical sickness. There is something deranged about the whole environmental-pollution-population crusade — something that is deeply hostile not merely to technological advance (that much is obvious) but to the very fact of human life as the West has understood it. Our ecological cult with its primitive worship of untended nature, its preference for fish over people, suggests an interior malaise — expressed as a kind of death wish for Western man and his society. If the "energy crisis" is allowed to run on much longer that wish could very possibly be fulfilled.

WHAT DO YOU MEAN NO GAS?

Equally to be excluded, as the teaching authority of the church has frequently declared, is direct sterilization, whether perpetual or temporary, whether of the man or of the woman.

Similarly excluded is every action which, either in anticipation of the conjugal act, or in its accomplishment, or in the development of its natural consequences, proposes, whether as an end or as a means, to render procreation impossible.

EXERCISE 7

EVALUATING SOURCES

A critical thinker must always question his various sources of information. Historians, for example, usually distinguish between *primary sources* (eyewitness accounts) and *secondary sources* (writings based on primary or eyewitness accounts, or other secondary sources). Most textbooks are examples of secondary sources. A diary written by a Civil War veteran is one example of a primary source. In order to be a critical reader one must be able to recognize primary sources. However, this is not enough. Eyewitness accounts do not always provide accurate descriptions. Historians may find ten different eyewitness accounts of an event and all the accounts might interpret the event differently. Then they must decide which of these accounts provide the most objective and accurate interpretations.

Test your skill in evaluating by participating in the following exercise. Pretend you are living 2000 years in the future. Your teacher tells you to write an essay analyzing the energy crisis in the United States during the 1970's.

First, *underline* only those descriptions you feel would serve as a primary source for your essay. Second, *rank* only the underlined or primary sources assigning the number (1) to the most objective and accurate primary source, number (2) to the next most accurate and so on until the ranking is finished. Then discuss and compare your evaluations with other class members.

Assume that all of the following essays, articles, and books deal with the energy crisis of the 1970's.

____ 1. A book written by Michael McClosky, executive director of the Sierra Club, in 1975.

____ 2. A speech by President Nixon in 1973.

____ 3. A 1972 newspaper editorial in the **New York Times.**

____ 4. An article written by a French journalist in France in 1975.

____ 5. An essay by an American sociologist, written in 1968.

____ 6. A speech by the president of Standard Oil.

____ 7. The findings of a congressional committee, published in 1978.

____ 8. An analysis by an American historian, written in 1992.

____ 9. A report by the Electrical Utility Industry, published in 1978.

READING NUMBER 15

THE MYTH OF THE PEACEFUL ATOM

by Richard Curtis & Elizabeth Hogan*

Richard Curtis & Elizabeth Hogan are co-authors of the book **Perils of the Peaceful Atom: The Myth of Safe Nuclear Power Plants.**

As you read try to answer the following questions:

1. According to the authors, what is the myth of the peaceful use of atomic power?
2. What specific hazards do the authors discuss?
3. What alternatives do the authors suggest to the use of atomic energy as a major source of electricity production?

*Richard Curtis & Elizabeth Hogan, *The Myth of the Peaceful Atom*, **Natural History,** March, 1969, p. 6-14ff. Reprinted with permission from Richard Curtis & Elizabeth Hogan.

The belief is widespread that the nuclear reactors being built to generate electricity for our cities are safe, reliable, and pollution-free. But a rapidly growing number of physicists, biologists, engineers, public health officials, and even staff members of the Atomic Energy Commission itself — the government bureau responsible for regulation of this force — have been expressing serious misgivings about the planned proliferation of nuclear power plants. In fact, some have indicated that nuclear power, which Supreme Court Justices William O. Douglas and Hugo L. Black described as "the most deadly, the most dangerous process that man has ever conceived," represents the gravest pollution threat yet to our environment. . . .

The hazards of peaceful atomic power fall into two broad categories: the threat of violent, massive releases of radioactivity or that of slow, but deadly, seepage of harmful products into the environment. . . .

All of us are familiar with technological disasters that have occurred against fantastically high odds: the sinking of the "unsinkable" *Titanic*, or the November 9, 1965, "blackout" of the northeastern United States, for example. The latter happening illustrates how an "incredible" event can occur in the electric utility field, most experts agreeing that the chain of circumstances that brought it about was so improbable that the odds against it defy calculation. . . .

A disturbing number of reactor accidents have occurred — with sheer luck playing an important part in averting catastrophe — that seem to have been the product of incredible coincidences. On October 10, 1957, for instance, the Number One Pile (reactor) at the Windscale Works in England malfunctioned, spewing fission products over so much territory that authorities had to seize all milk and growing foodstuffs in a 400-square-mile area around the plant. A British report on the incident stated that *all* of the reactor's containment features had failed. And, closer to home, a meltdown of fuel in the Fermi reactor in Lagoona Beach, Michigan, in October, 1966, came within an ace of turning into a nuclear "runaway." An explosive release of radioactive materials was averted, but the failures of Fermi's safeguards made the event, in the words of Sheldon Novick in *Scientist and Citizen*, "a bit worse than the 'maximum credible accident.' " . . .

Among the many factors contributing to reactor accidents, the human element is the most difficult to quantify. And perhaps for that reason, it has been largely overlooked in the AEC's assessments of reactor safety. . . .

AEC annuals are full of reports of human negligence: 3,844 pounds of uranium hexafluoride lost owing to an error in opening a cylinder; a $220,000 fire in a reactor because of accidental tripping of valves by electricians during previous maintenance work; numerous vehicular accidents involving transport of nuclear materials. None of these accidents led to disaster, but who will warrant that, with the projected proliferation of power plants and satellite industries in the coming decade, a moment's misjudgment will not trigger a nightmare? Perhaps worse, the likelihood of sabotage has scarcely been weighed, despite a number of incidents and threats. . . .

While there is little doubt that American technology is the most refined on earth, there is ample reason to believe that it has more than met its match in the seemingly insurmountable problems posed by the peaceful atom. . . .

Most serious of all, perhaps, is that tomorrow's reactors are now slated for location in close proximity to population concentrations. . . .

Is it necessary to build atomic plants so big and so close? The answer has to do with economics. The larger a facility is, the lower the unit cost of construction and operation and the cheaper the electricity. The longer the fuel cycle, the fewer the expensive shutdowns while spent fuel assemblies are replaced. The closer the plant is to the consumer, the lower the cost of rights of way, power lines, and other transmission equipment. . . .

Some of the deepest concern about the size and location of atomic plants has been expressed by members of the AEC themselves. "The actual experience with reactors in general is still quite limited," said Harold Price, AEC's Director of Regulation, in 1967 congressional hearings, "and with large reactors of the type now being considered, it is nonexistent. Therefore, because there would be a large number of people close by and because of lack of experience, it is . . . a matter of judgment and prudence at present to locate reactors where the protection of distance will be present."

Price's statement is mild compared to that made in the same hearings by Nunzio J. Palladino, Chairman of the AEC's Advisory Committee on Reactor Safeguards for 1967, and Dr. David Okrent, former Chairman for 1966: "the ACRS believes that placing large nuclear reactors close to population centers will require considerable further improvements in safety, and that *none of the large power reactors now*

From the beginning, insurance companies have avoided nuclear plants like the plague. By law, there is a strict liability ceiling of $560 million per accident, of which the government will cover up to $490 million. Back in 1957, the AEC estimated property damage of $7 billion from one big accident when reactors were much smaller. If the insurance industry has not wanted the business at any price, the public has a right to know why.

Ralph Nader, *Evidence Growing That Nuclear Plants Unsafe*, **Minneapolis Tribune**, March 31, 1973.

under construction is considered suitable for location in metropolitan areas (our italics)."

The threat of a nuclear plant catastrophe constitutes only half of the double jeopardy in which atomic power has placed us. . . .

Technology for retaining all radioactive contaminants, is either unperfected or costly, and much material of low-level radioactivity is routinely released into the air or water at the reactor site. These releases are undertaken in such a way, we are told, as to insure dispersion or dilution sufficient to prevent any predictable human exposure above harmful levels. Thus, when atomic power advocates are asked about the dangers of contaminating the environment, they imply that the relatively small amounts of radioactive materials released under "planned" conditions are harmless.

This view is a myth.

In the first place, many waste radionuclides take an extra-ordinarily long time to decay. The half-life (the time it takes for half of an element's atoms to disintegrate through fission) of strontium-90, for instance, is more than 27 years. Thus, even though certain long-lived isotopes are widely dispersed in air or diluted in water, their radioactivity does not cease. It remains, and over a period of time accumulates. It is therefore not pertinent to talk about the safety of any single release of "hot" effluents into the environment. At issue, rather, is their duration and cumulative radioactivity.

Further, many radioactive elements taken into the body tend to build up in specific tissues and organs to which those isotopes are attracted, increasing by many times the exposure dosage in those local areas of the body. Iodine-131, for instance, seeks the thyroid gland; strontium-90 collects in the bones; cesium-137 accumulates in muscle. Many isotopes have long half-lives, some measurable in decades.

Two more factors controvert the view that carefully monitored releases of low-level radioactivity into the environment are not pernicious. First, there is apparently no radiation threshold below which harm is impossible. Any dose, however small, will take its toll of cell material, and that damage is irreversible. Second, it may take decades for organic damage, or generations for genetic damage, to manifest itself. In 1955, for example, two British doctors reported a case of skin cancer — ultimately fatal — that had taken forty-nine years to develop following fluoroscopic irradiation of a patient.

Still another problem has received inadequate attention. Man is by no means the only creature in whom radioactive isotopes concentrate. The dietary needs of all plant and animal life dictate intake of specific elements. These concentrate even in the lowest and most basic forms of life. They are then passed up food chains, from grass to cattle to milk to man, for example. As they progress up these chains, the concentrations often increase, sometimes by hundreds of thousands of times. . . .

That nuclear facilities are producing dangerous build-ups of radio-isotopes in our environment can be amply documented. . . .

That "low-level" waste is a grossly deceptive term is obvious. In his book **Living with the Atom**, author Ritchie Calder in 1962 described an "audit" of environmental radiation that he and his colleagues, meeting at a symposium in Chicago, drew up to assess then current and future amounts of radioactivity released into atmosphere and water. Speculations covered the period 1955-65, and because atomic power plants were few and small during that time, the figures are more significant in relation to the future. Tallying "planned releases" of radiation from such sources as commercial and test reactors, nuclear ships, uranium mills, plutonium factories, and fuel-reprocessing plants, Calder's group came to a most disquieting conclusion: "By the time we had added up all the curies which might predictably be released, by all those peaceful uses, into the environment, it came to about 13 million curies per annum." A "curie" is a standard unit of radioactivity whose lethality can be appreciated from the fact that one trillionth of one curie of radioactive gas per cubic meter of air in a uranium mine is ten times higher than the official maximum permissible dose.

Calder's figures did not include fallout due to bomb testing and similar experiments, nor did they take into account possible reactor or nuclear transportation accidents. Above all, they did not include possible escape of stored high-level radioactive wastes, the implications of which were awesome to contemplate: "what kept nagging us was the question of waste disposal and of the remaining radioactivity which must not get loose. We were told that the dangerous waste, which is kept in storage, amounted to 10,000 million curies. If you wanted to play 'the numbers game' as an irresponsible exercise, you could divide this by the population of the world and find that it is over 3 curies for every individual."

Exactly what does Calder mean by "the question of waste disposal"?

It has been estimated that a ton of spent fuel in reprocessing will produce from forty to several hundred gallons of waste. This substance is a violently lethal mixture of short- and long-lived isotopes. It would take five cubic miles of water to dilute the waste from just one ton of fuel to a safe concentration. Or, if we permitted it to decay naturally until it reached the safe level — and the word "safe" is used advisedly — just one of the isotopes, strontium-90, would still be damaging to life 1,000 years from now, when it will have only one seventeen-billionth of its current potency.

There is no known way to reduce the toxicity of these isotopes; they must decay naturally, meaning *virtually perpetual containment.* Unfortunately, mankind has exhibited little skill in perpetual creations, and procedures for handling radioactive wastes leave everything to be desired. Formerly dumped in the ocean, the most common practice today is to store the concentrates in large steel tanks shielded by earth and concrete. This method has been employed for some twenty years, and about 80 million gallons of waste are now in storage in about 200 tanks. This "liquor" generates so much heat it boils by itself for years. Most of the inventory in these caldrons is waste from weapons production, but within thirty years, the accumulation from commercial nuclear power will soar if we embark upon the expansion program now being promoted by the AEC. Dr.

Donald R. Chadwick, chief of the Division of Health of the U.S. Public Health Service, estimated in 1963 that the accumulated volume of waste material would come to two billion gallons by 1995.

It is not just the volume that fills one with sickening apprehension but the techniques of disposing of this material. David Lilienthal put his finger on the crux of the matter when he stated: "These hugh quantities of radioactive wastes must somehow be removed from the reactors, must — without mishap — be put into containers that will never rupture; then these vast quantities of poisonous stuff must be moved either to a burial ground or to reprocessing and concentration plants, handled again, and disposed of, by burial or otherwise, with a risk of human error at every step." Nor can it be stressed strongly enough that we are not discussing a brief danger period of days, months, or years. We are talking of periods "longer," in the words of AEC Commissioner Wilfred E. Johnson, "than the history of most governments that the world has seen."

Yet already there are many instances of the failure of storage facilities. An article in an AEC publication has cited nine cases of tank failure out of 183 tanks located in Washington, South Carolina, and Idaho. And a passage in the AEC's authorizing legislation for 1968 called for funding of $2,500,000 for the replacement of failed and failing tanks in Richland, Washington. "There is no assurance," concluded the passage, "that the need for new waste storage tanks can be forestalled." If this is the case after twenty years of storage experience, it is beyond belief that this burden will be borne without some storage failures for centuries in the future. Remember too, that these waste-holding"tank farms" are vulnerable to natural catastrophes such as earthquakes, and to man-made ones such as sabotage. . . .

The burden that radioactive wastes place on future generations is cruel and may prove intolerable. Physicist Joel A. Snow stated it well when he wrote in *Scientist and Citizen:* "Over periods of hundreds of years it is impossible to ensure that society will remain responsive to the problems created by the legacy of nuclear waste which we have left behind."

"Legacy" is indeed a gracious way of describing the reality of this situation, for at the very least we are saddling our children and their descendants with perpetual custodianship of our atomic refuse, and at worst may be dooming them to the same agonizing afflictions and deaths suffered by those who survived Hiroshima. Radiation has been positively linked to cancer, leukemia, brain damage, infant mortality, cataracts, sterility, genetic defects and mutations, and general shortening of life. . . .

What must be done to avert the perils of the peaceful atom? A number of plans have been put forward for stricter regulation of activities in the nuclear utility field, such as limiting the size of reactors or their proximity to population concentrations or building more safeguards. As sensible as these proposals appear on the surface, they fail to recognize a number of important realities: first, that such arrangements would probably be opposed by utility operators and the government due to their prohibitively high costs. Since our government seems to be committed to making atomic power plants competitive with conventionally fueled plants, and because businesses are in business for profit, it is hardly likely they would buy these answers. Second, the technical problems involved in containment of radioactivity have not been successfully overcome, and there is little likelihood they will be resolved in time to prevent immense and irrevocable harm to our environment. Third, the nature of business enterprise is unfortunately such that *perfect* policing of the atomic power industry is unachievable. As we have seen in the cases of other forms of pollution, the public spirit of men seeking profit from industrial processes does not always rise as high as the welfare of society requires. It is unwise to hope that stricter regulation would do the job.

What, then, is the answer? The only course may be to turn boldly away from atomic energy as a major source of electricity production, abandoning it as this nation has abandoned other costly but unsuccessful technological enterprises.

READING NUMBER 16

NUCLEAR POWER IS SAFE AND CLEAN

by Glen T. Seaborg*

> Mr. Seaborg, a professor of chemistry at the University of California, was Chairman of the U.S. Atomic Energy Commission from 1961 until 1971. He is a co-discoverer of plutonium and nuclear isotopes, and in 1951 was one of the recipients of a Nobel Prize in chemistry.

Mr. Seaborg is a former chairman of the U.S. Atomic Energy Commission.

As you read consider the following questions:

1. What arguments does the author present to support his claim that nuclear power plants are safe?
2. What observations does the author make about thermal pollution?

*Dr. Glenn T. Seaborg, **The Environment — And What to Do About It** (Washington: United States Atomic Energy Commission, 1969). pp. 12-20.

The first thing that comes to mind relating nuclear energy to the environment is the role of nuclear power for generating electricity. Anyone who has ever visited a nuclear power station is bound to be impressed with its clean and quiet operation. Anyone who is the least knowledgeable about the technology of such a plant knows that the growth of nuclear power will help abate air pollution, will help reduce traffic and noise in the area surrounding the power plant, and generally should make that area a much more attractive and healthier place to be. Unfortunately, we in the nuclear field have not been as effective as we might be in getting across these points and in counteracting some of the general misunderstanding and apprehension a large segment of the public still shares over the safety of nuclear plants.

This is even more unfortunate in light of a recent resurgence of anti-nuclear articles designed to alarm the public about the growth of nuclear power when it should be enlightened about it. Many of these articles use the effective propaganda technique known as "stacking the deck" — the technique of taking all the detrimental, isolated facts and information about a subject, misinterpreting other factual material, adding numerous statements — taken out of context — by authorities in the field, and placing all this material in a story that gives a completely one-sided viewpoint. Specifically, every fact and statement in such a story may be true, while the article as a whole, and the conclusion it draws, may be invalid and misleading. Such dishonesty is made more harmful by the fact that these articles are written as exposes and crusades in the public interest.

I am not going to repeat to this audience all the facts you know so well about the care and safety exercised in the nuclear field and about our excellent safety record. I only wish it were possible to take every citizen who is seriously concerned about nuclear safety and who questions the integrity of those in the nuclear field concerned with nuclear safety and personally guide him through the planning, licensing, construction and operation of a nuclear power station. I would like to show him the study and consideration involved in the siting of such a plant, the care and inspection that goes into the design and manufacture of every component of the system. I would like to have him go through the licensing procedure and hear the questions asked at a review meeting of the Advisory Committee on Reactor Safeguards (ACRS) — sometimes for nuclear engineers the nearest thing to a reenactment of the Spanish

In summary then, I think that the point to be made is: that we know a great deal about radiation and the risks of nuclear power; that these risks are consistent and probably somewhat less than the risks from fossil fuel and other generally acceptable risks. We do need further study on both the long- and short-term risks to the public and the environment from fossil fuels. However, the risks (on a short-term basis) from producing electric power from thermal systems now in use is well within those risks normally accepted by society.

J.R. Coleman, Assistant Director, Division of Environmental Health, Minnesota Department of Health, for the Annual Conference of the Minnesota Environmental Health Association.

Inquisition. I would like him to go through the training of reactor engineers and witness the checking out, daily operation and monitoring of a nuclear station. Then I would like to ask him to compare nuclear power with any other technology or technical operation in the country, and question whether any has been handled as competently, with as much integrity or with as much care and concern for the public safety and well-being.

I would extend this procedure to the handling of the waste products of reactor fuel — contained and safely stored so that they pose no pollution problem. And for those concerned about the future of this reactor waste, accumulating from the projected growth of nuclear power, I would show them the progress we have made in concentrating and solidifying this waste, so that in the years ahead this material will not pose any dangerous "legacy" as some have suggested. As some of you know, I am sure, the solidification process will allow us to reduce each 100 gallons of high activity waste to only one cubic foot of solid waste. Such solid waste can be safely stored in salt mines. There are 400,000 square miles of salt deposits underlying the United States. And it has been estimated that all the solid waste produced by nuclear reactors in the year 2000 would occupy less than 1 percent of the volume of salt now being mined each year.

All this, I think, indicates that control of the radiation effects of nuclear power on the environment is well in hand, and that we can with the growth of nuclear power expect cleaner air and other environmental advantages not usually associated with generating plants.

I recognize that currently there is concern over the fact that nuclear power requires significant amounts of cooling water and that this raises the question of what has unfortunately been called "thermal pollution." Let me just make a few points to try to clarify some misconceptions on this subject.

First of all, while it is known that water temperatures, and variations in those temperatures, do have effects on aquatic life and its ecology and should be a factor in the siting of *all* thermal power plants — conventional and nuclear — we still have a lot to learn about such effects. A great deal of research is going on in this field and much is being learned. My colleague on the Commission, Dr. Gerald F. Tape, covered this subject quite thoroughly in his talk about the Washington Section of the American Nuclear Society last December. Until more evidence is in on the many studies being conducted on thermal effects, any broad, sweeping statements or recommendations about them cannot be fully honest or in the best interest of the country as a whole. Aside from the fact that water temperature variations generally affect different species in different ways — as you know, warmer water seems to benefit some while it is harmful to others — there are additional factors to consider in siting power plants. These involve the size and flow of the entire water system, its seasonal climate, projections of industry and power growth along its banks, its other uses and sources of pollution and many other factors. It is because of all these many factors, varying from one water system to another, from one geographical area to another, that the AEC currently favors legislation that will see thermal effects regulated by state water authorities with the cooperation and guidance of the Federal Water Pollution Control Administration (FWPCA).

As in many other of our environmental problems, the thermal effects situation has also produced "maximum feasible misunderstanding." But contrary to some of the impressions that have been established recently there is no need for hysteria over "thermal pollution." There are many solutions to this problem — technical and administrative — and they will be used. We have barely begun to tap the scientific and technological ingenuity that can solve this problem. There are even indications that in some areas the waste heat from power plants might be turned into a valuable environmental asset — useful for agriculture, aquaculture and recreational purposes.

Let me also assure anyone who doubts it of the concern and interest of the AEC in protecting and upgrading the nation's waterways and preserving its wildlife. We have not, as one publication so regrettably intimated, entered into a conspiracy with the electric utilities to promote nuclear power at the cost of harming the environment. From my many years of contact with the electric utilities, I know how seriously they take their responsibilities in this area, just as we do. And when they apply for a license for a new nuclear power reactor we urge them to cooperate in exploring its environmental and ecological considerations with such organizations as the Fish and Wildlife Service, the FWPCA and appropriate State agencies that may wish to consider the siting of these plants. . . .

I will only say at this point that all that I have seen and heard, my total experience in the nuclear energy field for more than a quarter of a century and my association with others who have devoted their lives to this field, has given me the firm conviction that the environmental problems associated with nuclear energy are manageable. With good planning and continued dedicated work on the part of those in the nuclear field, our electric utilities and those Government agencies that regulate our Nation's power systems, we can have safe, clean and reliable nuclear power — as much of it as we will need.

Glenn T. Seaborg before the Senate and House Joint Committee on Atomic Energy, *Environmental Effects of Producing Electric Power*, October 29, 1969.

We know today how much we live in a world of our own making. More than anything else our new environmental relationships — both with nature and our man-made environments — tell us this. They thrust on us awesome new responsibilities. They ask many probing and sometimes painful questions. They ask us to choose between new alternatives and to accept new risks. They also fill many people with many doubts about the human race — about its powers, its limitations, its role on this earth and now even its destiny in this universe. It is natural then that many of us are apprehensive, that some yearn for simpler times, that many warn "go slow" and others cry "retreat."

But there is no turning back. We are experiencing the birth shock of being born into a brave new world — and the bravery must be ours. It cannot be false bravery — an empty bravado of a mankind flaunting his knowledge and technology. It must be a courage tempered with compassion, a quest for more knowledge and understanding and the humility that comes when one reaches a peak only to gaze out on new and unexplored horizons, and then moves on. Our new "discovery" of our environment — the demands it makes — must bring us this kind of courage. It must make us a new and better breed of man. I think it will.

EXERCISE 8

DISTINGUISHING BETWEEN STATEMENTS THAT ARE PROVABLE AND THOSE THAT ARE NOT

From various sources of information we are constantly confronted with statements and generalizations about social problems. In order to think clearly about these problems, it is useful if one can make a basic distinction between statements for which evidence can be found, and other statements which cannot be verified because evidence is not available, or the issue is so controversial that it cannot be definitely proved. Students should constantly be aware that social studies texts and other information often contain statements of a controversial nature. The following exercise is designed to allow you to experiment with statements that are provable and those that are not.

In each of the following questions indicate whether you believe it is provable (P), too controversial to be proved to everyone's satisfaction (C), or unprovable because of the lack of evidence (U).

P = Provable
C = Too Controversial
U = Unprovable

____ 1. Anyone who has ever visited a nuclear power station is bound to be impressed with its clean and quiet operation.

____ 2. Nuclear power represents the most deadly and dangerous process that man has ever conceived.

____ 3. Solid nuclear wastes can be safely stored in salt mines.

____ 4. Most radioactive wastes are not presently being stored in salt mines.

____ 5. Liquid radioactive wastes cannot be safely stored.

____ 6. There have been many instances of the failure of storage facilities for liquid radioactive wastes.

____ 7. There is no need for hysteria over "thermal pollution." There are many solutions to this problem.

____ 8. The AEC both regulates and promotes the use of atomic energy.

____ 9. The regulatory division of the AEC received one half of one percent of the total AEC budget.

____ 10. The AEC has exercised great care in promoting the safe use of atomic energy.

____ 11. Nuclear power plants have become a necessity in our modern society.

____ 12. Man must turn boldly away from atomic energy as a major source of electricity production, abandoning it as this nation has abandoned other costly but unsuccessful technological enterprises.

____ 13. Nuclear power plants are not a danger to nearby communities.

APPENDIX

BUSINESS ORGANIZATIONS AND GOVERNMENT AGENCIES PUBLISHING INFORMATION ABOUT THE PROGRESS MADE IN CURBING POLLUTION

Thomas F. Engelhardt, Esq.
Counsel for AEC Regulator Staff
U.S. Atomic Energy Commission
Washington, D.C. 20545

Mr. Stanley T. Robinson, Jr.
Office of the Secretary
U.S. Atomic Energy Commission
Washington, D.C. 20545

Atomic Industrial Forum
850 Third Avenue
New York, New York 10022

James P. Gleason, Esq.,
Alternate Chairman
205 Commonwealth Building
1625 K Street, N.W.
Washington, D.C. 20006

Valentine B. Deale, Esq.,
Chairman
Atomic Safety and
Licensing Board
Suite 504
1001 Connecticut Avenue
Washington, D.C. 20036

Aigie A. Wells, Esq.
Chairman, Atomic Safety and
Licensing Board Panel
U.S. Atomic Energy Commission
Washington, D.C. 20545

Automobile Manufacturers
Association
1619 Massachusetts Avenue N.W.
Washington, D.C.

Dept. of Health Education and
Welfare
Washington, D.C.

Division of Public Information
U.S. Atomic Energy Commission
Washington, D.C. 20545

National Automobile Dealers Association
2000 K Street, N.W.
Washington, D.C.

J. Peter Koop
Information Supervisor
Northern States Power Company
414 Nicollet Mall
Minneapolis, Minnesota 55401

ECOLOGY ORGANIZATIONS

Environment Magazine
438 N. Skinker
St. Louis, Missouri 63130

Friends of the Earth (F.O.E.)
30 East 42nd Street
New York, N.Y. 10017
Telephone: 212-687-7847

The Izaak Walton League of America
Glenview, Ill. 60025
Telephone: 312-724-3880

League of Conservation Voters
917 15th Street, N.W.
Washington, D.C. 20005
Telephone: 202-638-2525

The League of Women Voters
1730 M Street, N.W.
Washington, D.C. 20036
Telephone: 202-296-1770

MECCA — Minnesota Environmental Control Citizens Association
Central Manor
26 East Exchange Street
St. Paul, Minn. 55101
Telephone: 612-222-2998

National Audubon Society
1130 5th Avenue
New York, N.Y. 10028
Telephone: 212-369-2100

National Wildlife Federation
1412 — 16th Street, N.W.
Washington, D.C. 20036

The Nature Conservancy
1522 K Street, N.W.
Washington, D.C. 20005
Telephone: 202-223-4710

Planned Parenthood-World Population
515 Madison Avenue
New York,, N.Y. 10022
Telephone: 212-752-2100

Sierra Club
1050 Mills Tower
San Francisco, Calif. 94104
Telephone: 415-981-8634

The Wilderness Society
729 15th Street N.W.
Washington, D.C. 20005
Telephone: 202-347-4132

Zero Population Growth
367 State Street
Los Altos, Calif. 94022

ACKNOWLEDGMENTS

illustration and picture credits

Page	
1	Justus in the **Minneapolis Star**. Reprinted with permission from the **Minneapolis Star**.
4	"Courtesy, the Washington (D.C.) Star-News".
11	Interpress film "attention".
15	Justus in the **Minneapolis Star**. Reprinted with permission from the **Minneapolis Star**.
26	Justus in the **Minneapolis Star**. Reprinted with permission from the **Minneapolis Star**.
32	Los Angeles Time Syndicate.
39	Justus in the **Minneapolis Star.** Reprinted with permission from the **Minneapolis Star**.
45	Washington Post — from the Herblock Gallery (Simon & Schuster, 1968).
52	Justus in the **Minneapolis Star**. Reprinted with permission from the **Minneapolis Star**.
56	Justus in the **Minneapolis Star**. Reprinted with permission from the **Minneapolis Star**.
60	Publisher's Hall Syndicate.
65	Justus in the **Minneapolis Star**. Reprinted with permission from the **Minneapolis Star**.
69	©1972 Chicago Daily News. By Fischetti.
77	Barnett in the **Indianapolis News**. Reprinted with permission from the **Indianapolis News.**
82	Reprinted with permission from the **Richmond News Leader.**
89	Barnett in the **Indianapolis News**. Reprinted with permission from the **Indianapolis News.**

meet the editors

GARY E. McCUEN, currently a social studies teacher at Eisenhower Senior High School in Hopkins, Minnesota, received his A.B. in history from Ripon College, and has an M.S.T. degree in history which he received from Wisconsin State University in Eau Claire, Wisconsin.

DAVID L. BENDER is a history graduate from the University of Minnesota. He also has an M.A. in government from St. Mary's University in San Antonio, Texas. He has taught social problems at the high school level and is currently working on additional volumes for the Opposing Viewpoints Series.